Challenges and Paradigms in Applied Robust Control

Challenges and Paradigms in Applied Robust Control

Editor

Beatrice Adamsen

scitus
academics

Challenges and Paradigms in Applied Robust Control
Edited by **Beatrice Adamsen**

ISBN: 978-1-68117-274-3
Library of Congress Control Number: 2016937445

www.scitusacademics.com
Box No. 4766, 616 Corporate Way,
Suite 2, Valley Cottage,
NY 10989

Printed in United States of America on Acid Free Paper

Preface

Robust control is a control theory branch that explicitly addresses uncertainty in its approach to controller design. Robust control methods are designed to work properly if some parameters or disturbances are found. In the presence of limited modeling errors, robust methods aim to achieve robust performance and/or stability. Unlike an adaptive control policy, a robust control policy is static; instead of adapting to variation measurements, the controller is designed to work on the assumption that certain variables are unknown but limited. Robust control systems often include advanced topologies involving multiple feedback loops and feed-forward paths. The control laws can be represented by the high order transfer functions required to perform the desired disruption rejection performance simultaneously with robust loop operation. In the late 1970s and early 1980s, the theory of robust control began and soon developed a number of techniques to deal with limited system uncertainty. The H-infinity loop shaping is probably the most important example of a robust control technique. In the area of robust control, there have been many recent advances.

The main goal of this book is to address important challenges and paradigms in the field of robust design and implementation of controls. The book highlights recent progress in the field and encourages and motivates new themes and solutions in the field of robust control.

Table of Contents

CHAPTER 1

Attitude Control of a Flexible Satellite by Using Robust Control Design Methods

Reza Mohsenipour[1], Hossein Nemati[2], Mehrzad Nasirian[1], Abdolreza Kashani Nia[1]

[1]Malekashtar University of Technology, Tehran, Iran

[2]East Tehran (QiamDasht) Branch, Islamic Azad University, Tehran, Iran

ABSTRACT

The increase of satellite's dimensions has caused flexibility and formation of uncertainty in their model. This is because of space missions being more complex and using light moving structures in satellites. Satellites are also encountered with various circumferential disturbance torques. This uncertainty in model and disturbance torques will cause undesirable performance of satellites' attitude control system. So, for attitude control of these satellites, those methods should be used which are robust to uncertainty of the plant's model and can reject the effects of disturbances and the measurement noise. One of these methods is the robust control design method. But, because of pole's place of these satellite's dynamics equations, the designing procedure of robust control will face difficulties. In this paper, by using an internal feedback as a novel idea, the satellite's dynamics equations are changed in such a way that the poles will be placed in proper locations. Then, for these new equations, by regarding the effects of flexibility as uncertainty and considering the uncertainty in inertia matrix of satellite, an H_∞ controller has been designed and for better performance, a μ-controller has been improved. Afterwards, these two controllers are analyzed and compared for the original dynamic equations, not for the modified ones. Also, for comparison, a classic controller has been also designed for the original plant and eventually all these three controllers are compared with each other.

KEYWORDS

Attitude Control of Satellite; Robust Control; Flexible. Satellite; Uncertainty

1. INTRODUCTION

Designing satellites with various missions such as climatic, military, geology and astrological missions will cause the payloads being increased and as a result, the increase of their dimensions, weight and consumed power. So, for more energy absorption, the effective section surface of satellite should be increased to installing more solar panels. On the other side, the existent limitations on satellites' launch will cause the restriction on their volume and weight. To decrease the volume of satellites, they are designed as a concentrated structure with some supplemental parts which are fastened before launch and are opened after settlement in orbit, and to decrease the weight of satellites, the light materials are used in designing structures. The whole of these factors; means light weight, low volume and large section surface will cause the flexibility of satellites' structure. In this case, preserving correct direction of main body and flexible parts encounter with many challenges. According to these realities, many theoretical researches have been done to identify and control flexible structures [1,2].

In the past three decades, the flexible satellites which are known as big spatial structures in some articles are considered a lot. In some NASA reports, the effect of satellites' flexibility in attitude control system, as unusual acts, has been mentioned [3]. More researches on this issue have specified that the reason of this strange act is the flexibility of the structure which will be intensified in some cases by attitude control system [4].

Before 70's decade, the attitude control and stabilization systems of satellites were designed according to dynamic modeling of rigid bodies and Single-Input SingleOutput (SISO) controllers. Along with the development of spatial sciences in the late 70's, big satellites which have flexible parts and include many sensors and actuators were considered. So, the need for using complex control laws and Multi-Input Multi-Output (MIMO) control systems for satellites were found to be essential [5]. For high-accuracy performance in pointing, three-axis attitude control will be used for satellites which lead to a MIMO control system. Generally, satellites are exposed to various circumferential disturbances such as solar pressure torques, magnetic torques, aerodynamic torques and celestial dust. Moreover, in most cases the inertia matrix of satellite isn't specified exactly. So, designing the control system for flexible satellites will be encountered with a lot of problems [6]. Three main issues in designing attitude control of satellite are the space environmental disturbance, uncertainty in flexible structure and inertia matrix of satellite as well as control input constraints [7].

A review of using robust control for satellites is as follows. In [8], designing of an optimal low-order robust control with alpha-shift for attitude control of a spacecraft with solar electric propulsion has been discussed. In this article, a new method of block-shift for replacing different sets of closed-loop eigenvalues by different values has been introduced. In [9], a mixed H_2/H_∞ controller is designed which considers both stability robustness and root-mean-square performance and is of particular interest for attitude control of microsatellites. The mixed H_2/H_∞ controller shows a remarkable capability of achieving a balanced tradeoff between H_2 and H_∞ performances. The authors in [10] describe modeling of a flexible structure and three-axis controller design process. Also, they demonstrate the suitable performance of this design in maneuverability by applying it to a large flexible spacecraft model. In [7], a mixed H_2/H_∞ output-feedback controller has been proposed for microsatellite attitude control with pole placement constraints against the internal uncertainty of momentof-inertia variation and space environmental disturbances. The simulation results demonstrate that the presented mixed H_2/H_∞ control system is robust stable and optimal in the sense of H_2 norm, and has good steadystate and dynamic performance against parameter uncertainties and various disturbances for the microsatellite attitude control system. In [11], an optimization-based approach is proposed for the robustness analysis of an attitude and orbit control system (AOCS) for flexible satellites.

In this paper, it is intended to design robust attitude control for a flexible satellite with a desirable accuracy and performance against the uncertainty in the satellite's model and inertia matrix and with the existence of control input constraints and in the presence of environmental disturbances and measurement noise. The rest of this paper is as follows. In Section 2, the dynamic equations of a flexible satellite's attitude and the actuator of reaction wheel has been derived and the disturbances of space environment are discussed. Since the desirable attitude of satellite' tracking has small angles around zero, the extracted equations of the satellite will be linearized and combined with the equations of reaction wheel. In Section 3, it is shown that the original linearized equations of the satellite causes some problem in the robust control design procedure. So by using two unit feedbacks, its poles will be placed in a more proper place and new equations will be obtained. Also, in this section a nominal model for the satellite and needed weight functions for designing robust controller will be selected. In Section 4, a classical controller, based on Euler angels errors, for the original equations and an H_∞ controller, by using the new idea of internal feedback, for the modified equations will be designed. But, neither will achieve the desired robust performance for the original plant. Afterwards, to improve the performance of attitude control, a μ-controller for the modified equations will be designed to ensure the robust performance for the originnal plant. Using the obtained results for all three controllers, the paper will be concluded in Section 5.

2. SYSTEM AND DISTURBANCES MODELING

2.1. Satellite's Attitude and Reaction Wheel Modeling

The kinematic equations of satellite's attitude are as below,

$$\begin{aligned}\dot{\phi} &= p+[q\sin\phi+r\cos\phi]\tan\theta \\ \dot{\theta} &= q\cos\phi-r\sin\phi \\ \dot{\psi} &= [q\sin\phi+r\cos\phi]\sec\theta\end{aligned} \quad (1)$$

in which ϕ, θ and ψ are roll, pitch and yaw and $\boldsymbol{\omega}_{RB}^{B}=[p,q,r]^{\mathrm{T}}$ is the angular velocity vector of orbital reference frame with respect to body frame stated in body frame. According to the definition, the orbit reference frame's center conforms to satellite's center of mass, Z_R axis is towards the earth center, X_R axis is in direction of satellite's velocity and Y_R axis is vertical to plane of orbit so that transform the frame to a right-hand orthogonal frame. The body frame will be selected in a way that its axes conform to main axes of inertia. Vector $\boldsymbol{\omega}_{RB}^{B}$ satisfies the equation

$$\begin{aligned}\boldsymbol{\omega} &= \boldsymbol{\omega}_{BR}^{B}+\boldsymbol{\omega}_{RI}^{B} \\ &= \boldsymbol{\omega}_{BR}^{B}-\omega_0\begin{bmatrix} C\theta\cdot S\psi \\ C\phi\cdot C\psi+S\phi\cdot S\theta\cdot S\psi \\ -S\phi\cdot C\psi+C\phi\cdot S\theta\cdot S\psi\end{bmatrix}\end{aligned} \quad (2)$$

in which ω_0 is the angular velocity of orbit, C and S are accordingly for $\cos$ and $\sin$ and $\boldsymbol{\omega}=[\omega_x,\omega_y,\omega_z]^{\mathrm{T}}$ is the angular velocity of body frame with respect to inertial frame which will be obtained from Euler's moment equations. Assuming that body frame axes conforming to main inertial axes, Euler's moment equations is as

$$\boldsymbol{T}_d + \boldsymbol{T}_G = \begin{bmatrix} \dot{h}_x + \dot{h}_{wx} + \left(\omega_y h_z - \omega_z h_y\right) + \left(\omega_y h_{wz} - \omega_z h_{wy}\right) \\ \dot{h}_y + \dot{h}_{wy} + \left(\omega_z h_x - \omega_x h_z\right) + \left(\omega_z h_{wx} - \omega_x h_{wz}\right) \\ \dot{h}_z + \dot{h}_{wz} + \left(\omega_x h_y - \omega_y h_x\right) + \left(\omega_x h_{wy} - \omega_y h_{wx}\right) \end{bmatrix} \tag{3}$$

in which $\boldsymbol{T}_d$ is disturbance moment, $\boldsymbol{h}$ is angular momentum vector of satellite's rigid body, $\boldsymbol{h}_w$ is the angular momentum vector of reaction wheel and $\boldsymbol{T}_G$ is the gravity gradient moment which will be obtained by Equation (4).

$$\boldsymbol{T}_G = \begin{bmatrix} 1.5\omega_0^2 \left(I_z - I_y\right) \sin 2\phi \cos^2\theta \\ 1.5\omega_0^2 \left(I_z - I_x\right) \sin 2\theta \cos\phi \\ 1.5\omega_0^2 \left(I_x - I_y\right) \sin 2\theta \sin\phi \end{bmatrix} \tag{4}$$

Equations (1) to (4) form the dynamic nonlinear equations of satellite's attitude [12].

Here, to produce the moment, three reaction wheels will be used in direction of body axes. The block diagram of reaction wheel to control the attitude of satellite is as **Figure 1**. In this figure u is the output of controller and $\dot{h}_w$ is the exerted torque to satellite in direction of related axis [12].

Transfer function of **Figure 1** is approximately as follows [12].

$$\frac{\dot{h}_w}{u}(s) \approx 1 \tag{5}$$

So, the combination of the satellite's equations and the actuators is approximately the same satellite's equations, and the controller for satellite's equations will be designed apart from reaction wheel equations. Since the desirable angle which the satellite should track is around the zero angles and also to use the concepts of linear control and simplifying the analysis, the Equations (1)- (4) can be linearized. By linearizing these equations around zero point, linear state space form of dynamic equations of satellite's attitude will be obtained as

$$\begin{cases} \dot{X} = AX + Bu \\ y = CX + Du \end{cases} \tag{6}$$

in which $\boldsymbol{X} = \left[\phi, \theta, \psi, \dot{\phi}, \dot{\theta}, \dot{\psi}, h_{wx}, h_{wy}, h_{wz}\right]^{\mathrm{T}}$,

$\boldsymbol{u} = \left[\dot{h}_{wx}, \dot{h}_{wy}, \dot{h}_{wz}\right]^{\mathrm{T}}$ and

$$
\boldsymbol{A} = \begin{bmatrix}
0 & 0 & 0 & 1 & 0 & 0 & 0 & 0 & 0 \\
0 & 0 & 0 & 0 & 1 & 0 & 0 & 0 & 0 \\
0 & 0 & 0 & 0 & 0 & 1 & 0 & 0 & 0 \\
\frac{-4\omega_0^2}{I_x}\left(I_y - I_z\right) & 0 & 0 & 0 & 0 & \frac{\omega_0}{I_x}\left(I_x - I_y + I_z\right) & 0 & 0 & \frac{\omega_0}{I_x} \\
0 & \frac{-3\omega_0^2}{I_y}\left(I_x - I_z\right) & 0 & 0 & 0 & 0 & 0 & 0 & 0 \\
0 & 0 & \frac{-\omega_0^2}{I_z}\left(I_y - I_x\right) & \frac{-\omega_0}{I_z}\left(I_x + I_z - I_y\right) & 0 & 0 & \frac{-\omega_0}{I_z} & 0 & 0 \\
0 & 0 & 0 & 0 & 0 & 0 & 0 & 0 & 0 \\
0 & 0 & 0 & 0 & 0 & 0 & 0 & 0 & 0 \\
0 & 0 & 0 & 0 & 0 & 0 & 0 & 0 & 0
\end{bmatrix}
$$

$$
\boldsymbol{B} = \begin{bmatrix}
0 & 0 & 0 \\
0 & 0 & 0 \\
0 & 0 & 0 \\
-\frac{1}{I_x} & 0 & 0 \\
0 & -\frac{1}{I_y} & 0 \\
0 & 0 & -\frac{1}{I_z} \\
1 & 0 & 0 \\
0 & 1 & 0 \\
0 & 0 & 1
\end{bmatrix}, \quad
\boldsymbol{C} = \begin{bmatrix}
1 & 0 & 0 & 0 & 0 & 0 & 0 & 0 & 0 \\
0 & 1 & 0 & 0 & 0 & 0 & 0 & 0 & 0 \\
0 & 0 & 1 & 0 & 0 & 0 & 0 & 0 & 0
\end{bmatrix}, \quad \boldsymbol{D} = 0
$$

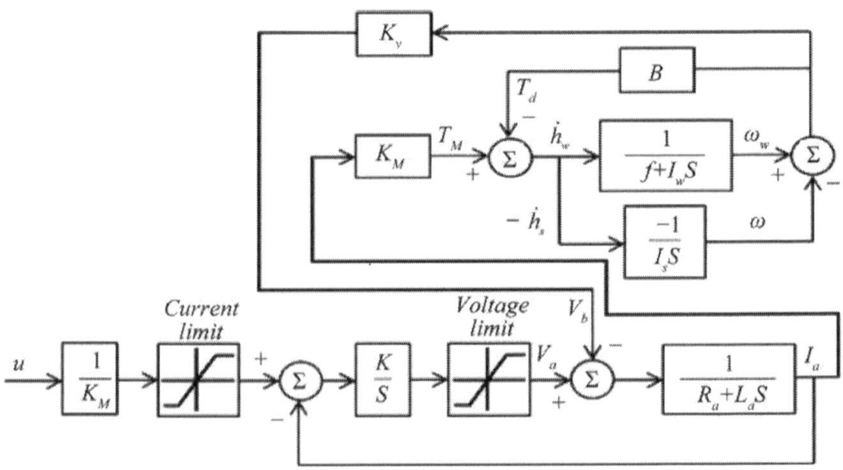

Figure 1. The block diagram of reaction wheel to control the attitude of satellite [12].

Satellite's configuration is as **Figure 2**. The system response of satellite' attitude with given specifications in **Table 1**, to one step with amplitude 10^{-5} N·m has been shown in **Figure 3**. As it could be seen, the system is unstable in each three channels of roll, pitch and yaw.

2.2. Disturbances Modeling

One of the key issues in attitude control of satellites is the environment disturbances which deflect the attitude of satellite from its desirable state and ignoring these disturbances make the designed controller completely useless.

Magnetic torque: a magnetic torque will affect on satellite's configure because of earth's magnetic field and satellite's magnetic materials. This torque will be obtained by

$$\boldsymbol{T}_m = \boldsymbol{M} \times \boldsymbol{B} \tag{7}$$

in which $\boldsymbol{M}$ is the residual magnetic moments of the satellite caused by permanent and induced magnetism and the satellite-generated current loops and B is the geocentric magnetic flux density. Vector B will be obtained, in orbit reference system, by

$$\boldsymbol{B} = \frac{\mu_f}{|\boldsymbol{r}|^3} \begin{bmatrix} \cos\omega_0 t \sin i_m \\ \cos i_m \\ 2\sin\omega_0 t \cos i_m \end{bmatrix} \tag{8}$$

in which $\mu_f = 7.9\times10^{15}$ Wb·m and $\boldsymbol{r}$ is the beam of satellite's orbit and i_m is the angle of satellite's orbit with respect to geomagnetic equator [13].

Aerodynamic torque: because of satellite's movement in upper atmosphere of earth, an aerodynamic torque will affect on satellite. This torque can be obtained as

$$\boldsymbol{T}_a = \frac{1}{2}\rho|\boldsymbol{v}|^2 C_d A_a \left(\boldsymbol{u}_a \times \boldsymbol{s}_{cp}\right) \tag{9}$$

in which ρ is the atmospheric density, $\boldsymbol{v}$ is the velocity of satellite, C_d is the drag coefficient, $\boldsymbol{u}_a$ is the unit vector along velocity direction, A_a is the area of vertical surface on $\boldsymbol{u}_a$ and $\boldsymbol{s}_{cp}$ is a vector from satellite' center of mass to center of pressure [9].

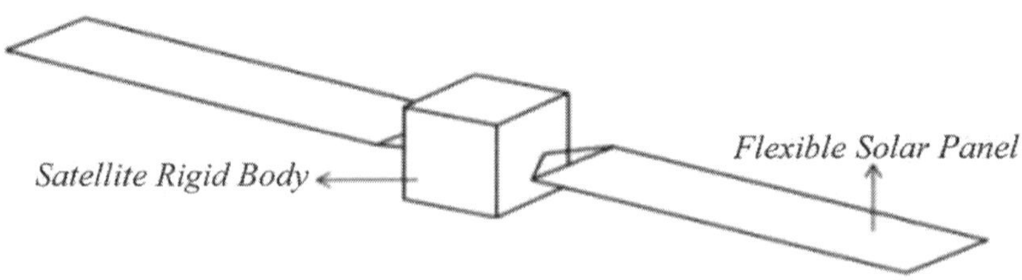

Figure 2. The satellite's configuration.

Table 1. Satellite physical and orbital characteristics.

value	Parameter
0.75 × 0.75 ×0.75	Size of satellite rigid body (m)
3 × 0.75	Size of a solar panel (m)
15	Weight (kg)
[1 4 2]	Moment of inertia (kg·m^2)
207.2033	Mean anomaly (degree)
0.0143013	Eccentricity
55.5535	Inclination (degree)
153.6389	Argument of perigee (degree)
315.2281	Right ascension of ascending node (degree)

Solar torque: because of the contact of solar radiation particles with satellite, a disturbance force will exert on satellite. This force can be obtained approximately by

$$\boldsymbol{F}_s = \frac{1367}{c} A_s (1+q) \cos \gamma \boldsymbol{u}_s \tag{10}$$

in which c is the velocity of light, A_s is the area opposed to sun radiation, q is the reaction coefficient, γ is the radiation angle, $\boldsymbol{u}_s$ is the unit vector along sun radiation direction and $\boldsymbol{C}_{cp}$ is the distance from the satellite's center of mass to the solar pressure center [12].

Now, by considering the dynamic equations of satellite's attitude and reaction wheel, the object is to design an attitude control system as shown in **Figure 4**.

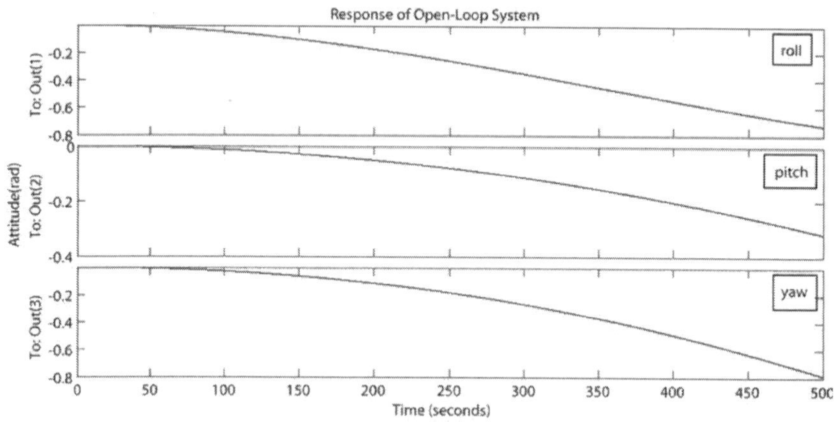

Figure 3. Response of open-loop system.

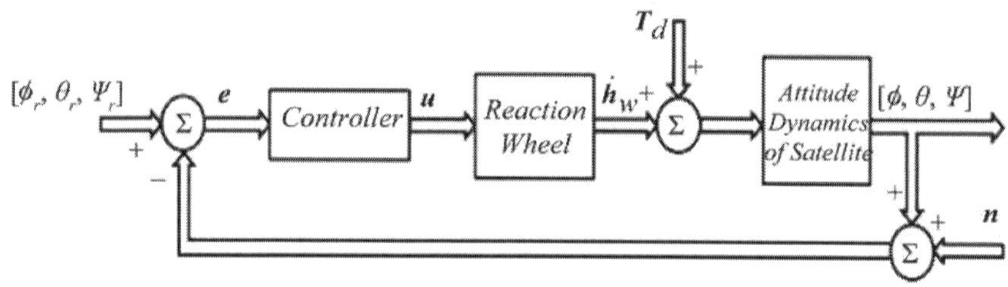

Figure 4. The block diagram of satellite attitude control system.

3. SELECTION OF NOMINAL MODEL AND NEEDED WEIGHT FUNCTIONS

The modeling of flexible satellite will be done in a way that a nominal model $P_0(s)$ will be selected and flexibility effects will be considered as uncertainty around the nominal model. These uncertainties lead to the perturbed models $P(s)$. Equations (6) with nominal amounts of **Table 1**will be considered as nominal model. To consider flexibility effects, it is assumed that the inertia moments of satellite have 30 percentage of uncertainty. Since the system has unmodelled flexible parts, the uncertainty will be considered as multiplicative unstructured. For a SISO system, the perturbed model structure of $P(s)$ based on nominal model of $P_0(s)$, multiplicative unstructured uncertainty block of $\Delta(s)$ and uncertainty weight function of $W(s)$ is as

$$P(s)=\left[1+\Delta(s)W(s)\right]P_0(s) \tag{11}$$

which its block diagram is shown in **Figure 5**.

The weight function of $W(s)$ should be selected in a way that relation

$$\forall\omega:\left|w(s)\right|\geq\left|\frac{P(s)}{P_0(s)}-1\right| \tag{12}$$

by condition $\|\Delta\|_\infty\leq 1$ will be established for all perturbed models of $P(s)$ [14]. The transfer function matrix of linearized system is obtained as follows

$$P_0(s)=\begin{bmatrix} P_{0_{11}}(s) & 0 & P_{0_{13}}(s) \\ 0 & P_{0_{22}}(s) & 0 \\ P_{0_{31}}(s) & 0 & P_{0_{33}}(s) \end{bmatrix} \tag{13}$$

The satellite' attitude is a MIMO system, so for each element of this matrix, a weight function of $w_{ij}(s)$ should be selected in such a way that, it satisfies the relation (12) for i, j = 1, 2, 3. For elements with zero value, the uncertainty weight function could be equal to 1. Bode diagram of $\left|P_{ij}(s)/P_{ij_0}(s)-1\right|$ for the non-zero elements of Equation (13) has been shown in **Figure 6**. It could be seen that except for the $\left|P_{22}(s)/P_{22_0}(s)-1\right|$, other figures have big overshoots because of placement of some poles on imaginary axis. So, the weight functions of $w_{ij}(s)$ corresponding to those which should satisfy the relation (12), have also same overshoots and this will cause the issue of robust controller design face difficulties [14]. The poles of non-zero elements of Equation (13) are as

$$p_1 = 0,\ p_{2,3} = \pm i\sqrt{1.508\times10^{-6}},\ p_{4,5} = \pm i\sqrt{9.308\times10^{-6}}$$

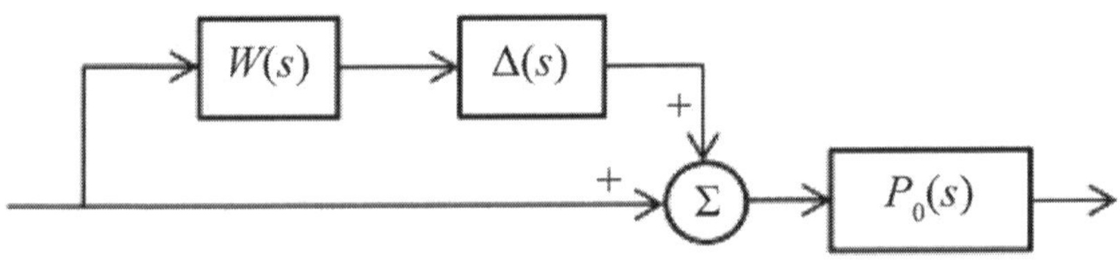

Figure 5. Block diagram of the perturbed model of P with multiplicative unstructured uncertainty.

One of the ways for elimination of this problem is displacement of poles slightly [8]. But, here the amounts of inertia moments of satellite are so that the system's poles which are on imaginary axis are very close to the origin and the problem won't be solved by this displacement, too.

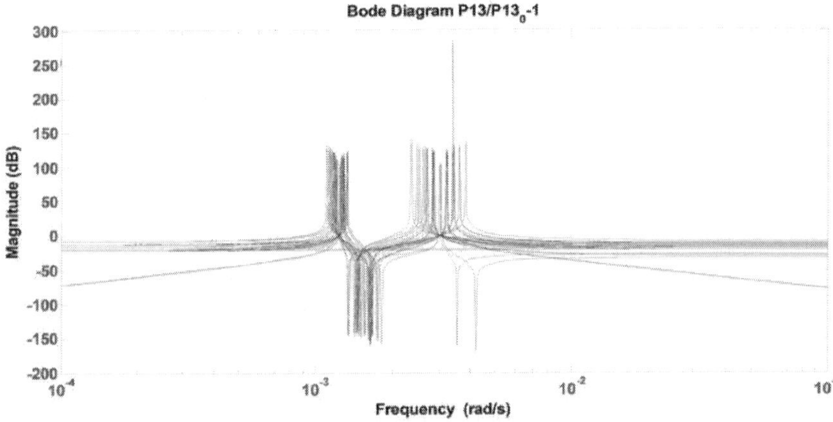
Bode Diagram P13/P13$_0$-1
Magnitude (dB)
Frequency (rad/s)

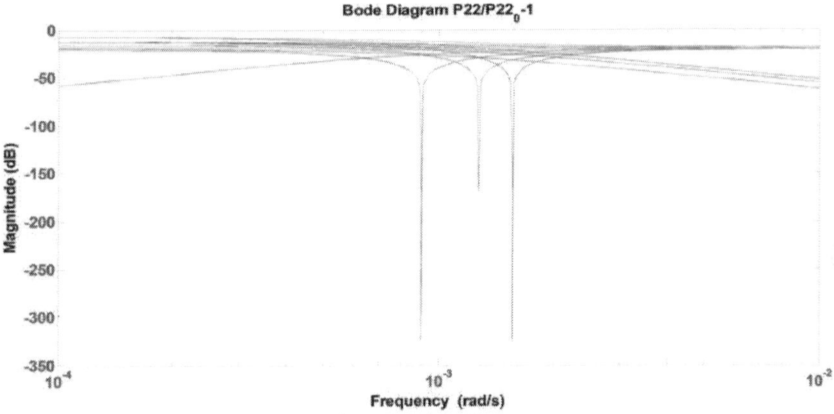
Bode Diagram P22/P22$_0$-1
Magnitude (dB)
Frequency (rad/s)

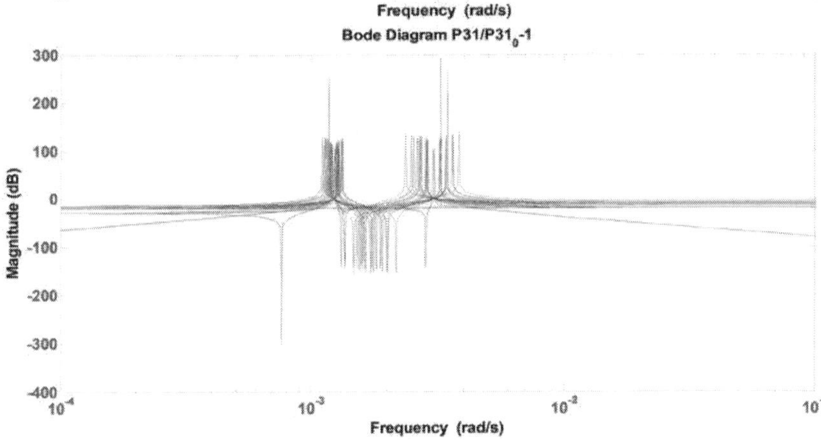
Bode Diagram P31/P31$_0$-1
Magnitude (dB)
Frequency (rad/s)

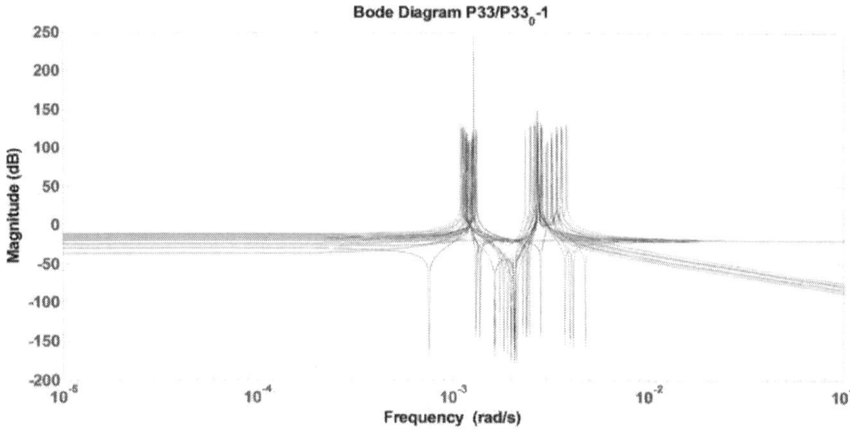

Figure 6. Bode diagram of $\left|P_{ij}(s)/P_{ij_0}(s)-1\right|$ for the non-zero elements of Equation (13).

For overcoming this problem, equations should be changed in a way that poles go sufficiently away from imaginary axis. With a new idea, according to **Figure 7**two unit internal feedbacks in the main nonlinear plant will be made from roll and yaw outputs to first and third inputs respectively. So, the state space's form of the modified linearized plant will be obtained as

$$\begin{cases} \dot{X} = A_{\text{new}} X + Bu' \\ y = CX + Du' \end{cases} \tag{14}$$

where

$$A_{new} = \begin{bmatrix} 0 & 0 & 0 & 1 & 0 & 0 & 0 & 0 & 0 \\ 0 & 0 & 0 & 0 & 1 & 0 & 0 & 0 & 0 \\ 0 & 0 & 0 & 0 & 0 & 1 & 0 & 0 & 0 \\ \frac{-4\omega_0^2\left(I_y - I_z\right)+1}{I_x} & 0 & 0 & 0 & 0 & \frac{\omega_0}{I_x}\left(I_x - I_y + I_z\right) & 0 & 0 & \frac{\omega_0}{I_x} \\ 0 & \frac{-3\omega_0^2}{I_y}\left(I_x - I_z\right) & 0 & 0 & 0 & 0 & 0 & 0 & 0 \\ 0 & 0 & \frac{-\omega_0^2\left(I_y - I_x\right)+1}{I_z} & \frac{-\omega_0}{I_z}\left(I_x - I_y + I_z\right) & 0 & 0 & \frac{-\omega_0}{I_z} & 0 & 0 \\ 0 & 0 & 0 & 0 & 0 & 0 & 0 & 0 & 0 \\ 0 & 0 & 0 & 0 & 0 & 0 & 0 & 0 & 0 \\ 0 & 0 & 0 & 0 & 0 & 0 & 0 & 0 & 0 \end{bmatrix}$$

and **Figure 4** will change to **Figure 8**.

The matrix of transfer function of the modified plant is as Equation (15).

$$\mathcal{P}_0(s) = \begin{bmatrix} \mathcal{P}_{0_{11}}(s) & 0 & \mathcal{P}_{0_{13}}(s) \\ 0 & \mathcal{P}_{0_{22}}(s) & 0 \\ \mathcal{P}_{0_{31}}(s) & 0 & \mathcal{P}_{0_{33}}(s) \end{bmatrix} \tag{15}$$

The poles of non-zero elements of Equation (15) are

$$p_1 = 0,\ p_{2,3} = \pm 1,\ p_{4,5} = \pm 0.7071.$$

Bode diagram of $\left|\mathcal{P}_{ij}(s)/\mathcal{P}_{ij_0}(s) - 1\right|$ for the non-zero elements of Equation (15) are shown in**Figure 9**. It can be seen that overshoots are eliminated. Now, weight functions of $w_{ij}(s)$ will be selected in a way that satisfies the Equation (12) for the modified plant. This is done and shown in **Figure 9**.

The condition of robust stability for a SISO system is

$$\left\| w(s)T(s) \right\|_\infty < 1 \tag{16}$$

in which $T(s)$ is the transfer function of desirable closed-loop system and $w(s)$ is the weight function of uncertainty [15]. Since the elements of Equation (15) are not reachable separately, the matrix of weight function will be selected as

$$W(s) = \begin{bmatrix} w_1(s) & 0 & 0 \\ 0 & w_2(s) & 0 \\ 0 & 0 & w_3(s) \end{bmatrix} \tag{17}$$

Where

$$\forall \omega : \left| w_i(s) \right| \geq \max_{j=1,2,3} \left| w_{ij}(s) \right| \tag{18}$$

So, weight functions will be obtained as Equations (19).

$$W_1(s) = 10^{-\frac{5}{20}} (s + 0.0025)^2 \Big/ (s + 0.8)^2$$

$$W_2(s) = 10^{-\frac{18}{20}} (S + 0.02) \Big/ (S + 0.0005)$$

$$W_3(s) = 10^{-\frac{3}{20}} (s + 0.0015)^2 \Big/ (s + 0.8)^2$$

The condition of nominal performance for a SISO system is as

$$\left\| w_s(s) s(s) \right\|_\infty < 1 \qquad (20)$$

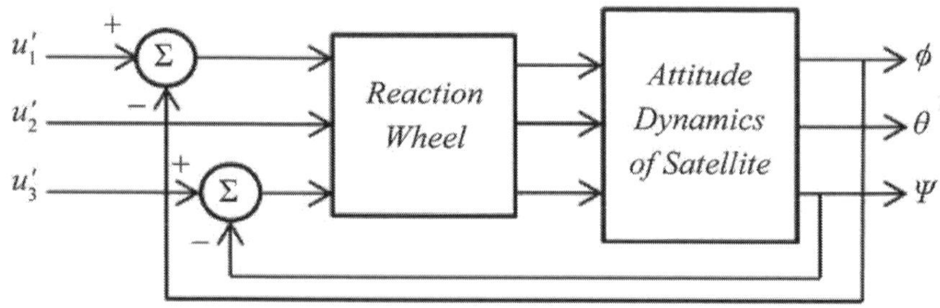

Figure 7. To make internal feedback in the main system.

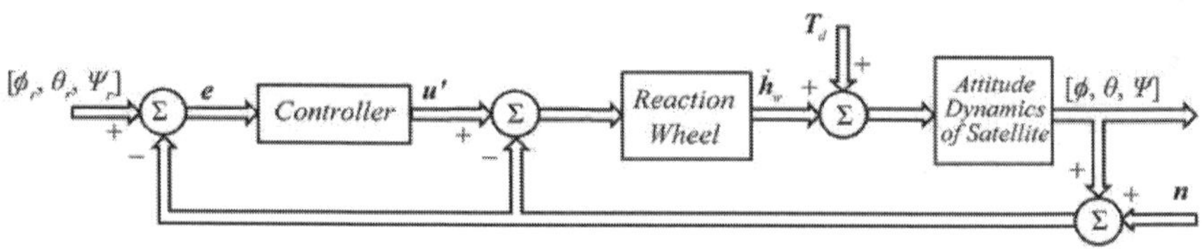

Figure 8. Block diagram of satellite attitude control with internal feedback.

in which $s(s)$ is transfer function of the error to reference input and $w_s(s)$ is the sensitivity weight function [15]. Here, it's desirable that the output of closed-loop system track a sinus route with minimum control effort and error less than 2°. So, the matrix of sensitivity weight function for the discussed system with trial and error will be selected as

$$W_s(s) = \begin{bmatrix} w_{s_1}(s) & 0 & 0 \\ 0 & w_{s_2}(s) & 0 \\ 0 & 0 & w_{s_3}(s) \end{bmatrix} \qquad (21)$$

Where

$$W_{s_i}(s) = \frac{S^2 + 1.5S + 2.25}{S(S+1.5)}, \quad i = 1,2,3 \tag{22}$$

For limiting the controller output, beside the weight functions of uncertainty and sensitivity, a related weight function will be selected as $w_U(s) = 0.001$.

4. CONTROLLER DESIGN

In this part, for the Equations of original plant (6), a classical controller, based on Euler angles errors, and for the Equations of the modified plant (14), two robust controllers H_∞ and μ will be designed. Eventually, the original nonlinear system, described by Equations (1)-(4), will be simulated with each of three controllers in closedloop form. In simulations, the inertia moments of satellites have 30 percent of uncertainty as shown in **Figure 10**. The simulation of the closed-loop system has done in two cases. In one case, the closed-loop system response is obtained to zero reference input and with $\boldsymbol{T}_d$ equal to disturbance pulse, which is shown in **Figure 11**. In the other case, the closed-loop system response to a reference sinus input with the amplitude of 15 degree in roll channel, 10 degree in pitch channel and 5 degree in yaw channel and 20 Hz frequency in the presence of space physical disturbances and measurement noise, is obtained to examine the performance of system in tracking. The RMS amount of attitude sensors' noise of and their error are considered equal to 0.01°.

4.1. Classical Controller

When a classical controller is designed for the equations of rigid body of a satellite, although it can be robust somehow against the uncertainty, but there isn't any guarantee to have a desirable performance for a practical plant. Here, for the original plant with Equations (6), a classical controller has been designed based on Euler angles' errors and the control laws are as Equations (23) with coefficients $K_x = 2.25$, $K_{xd} = 1.5$, $K_y = 9$, $K_{yd} = 6$, $K_z = 4.5$ and $K_{yz} = 3$ [12].

$$u_1 = K_x(\phi_r - \phi) + K_{xd}\dot{\phi}$$
$$u_2 = K_y(\theta_r - \theta) + K_{yd}\dot{\theta}$$
$$u_3 = K_z(\psi_r - \psi) + K_{yz}\dot{\psi} \qquad (23)$$

The closed-loop system response to disturbance pulse for this controller has drawn in **Figure 12**. Its performance in tracking and control effort has also been shown in Figures 13 and 14 respectively. Tracking error's norm is about 5°.

4.2. H_∞ Suboptimal Controller

Interconnections of system for designing H_∞ suboptimal controller, regarding to selected weight functions, are as **Figure 15**. In this figure, d is the disturbance and p_{in} and p_{out} are the input and output of uncertainty block, respectively. By considering **Figure 16**, the object is to find the controller $K(s)$ which satisfies Equation (24). In this equation, $G(s)$ is the generalized system [14].

$$\min_{K \text{ stabilising}} \left\| F_l(G,K) \right\|_\infty \qquad (24)$$

Designing H_∞ suboptimal controller could be done in MATLAB by using the hinsyn command. This command will design an H_∞

suboptimal controller for the open-loop system in **Figure 15** [16]. Here, by using this command, an H_∞ suboptimal controller with the degree of 22 is obtained.

For the closed-loop system analysis with the H_∞ suboptimal controller, from the aspect of robust stability, nominal performance and robust performance, the structure shown in **Figure 17**, will be used. In this figure, the uncertainty has been gathered in Δ block which is considered as Equation (25).

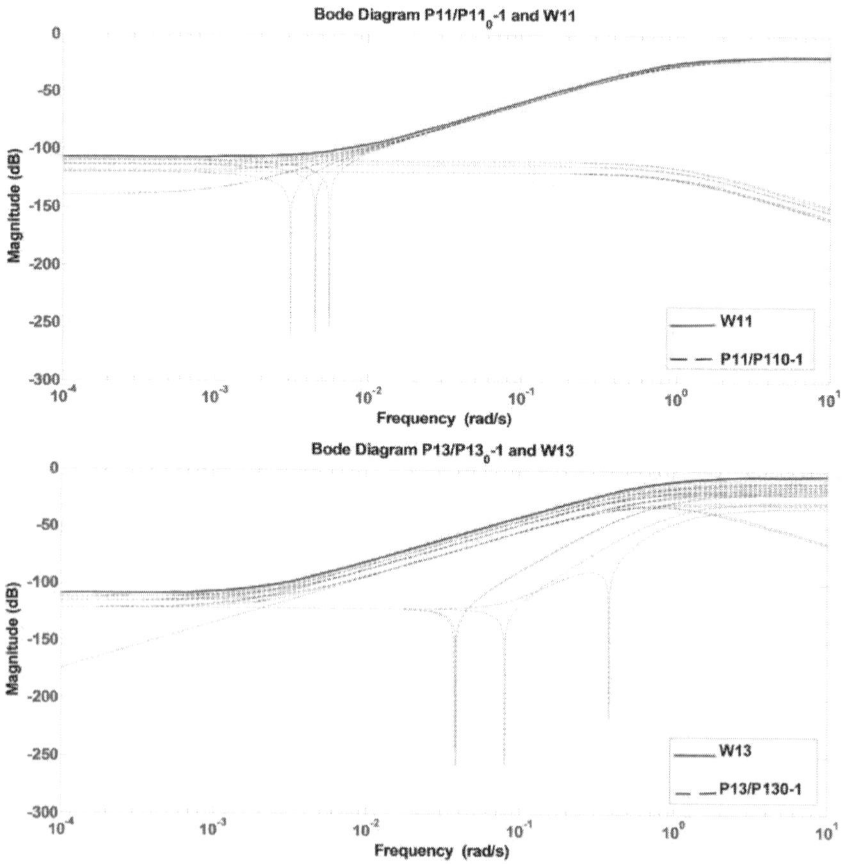

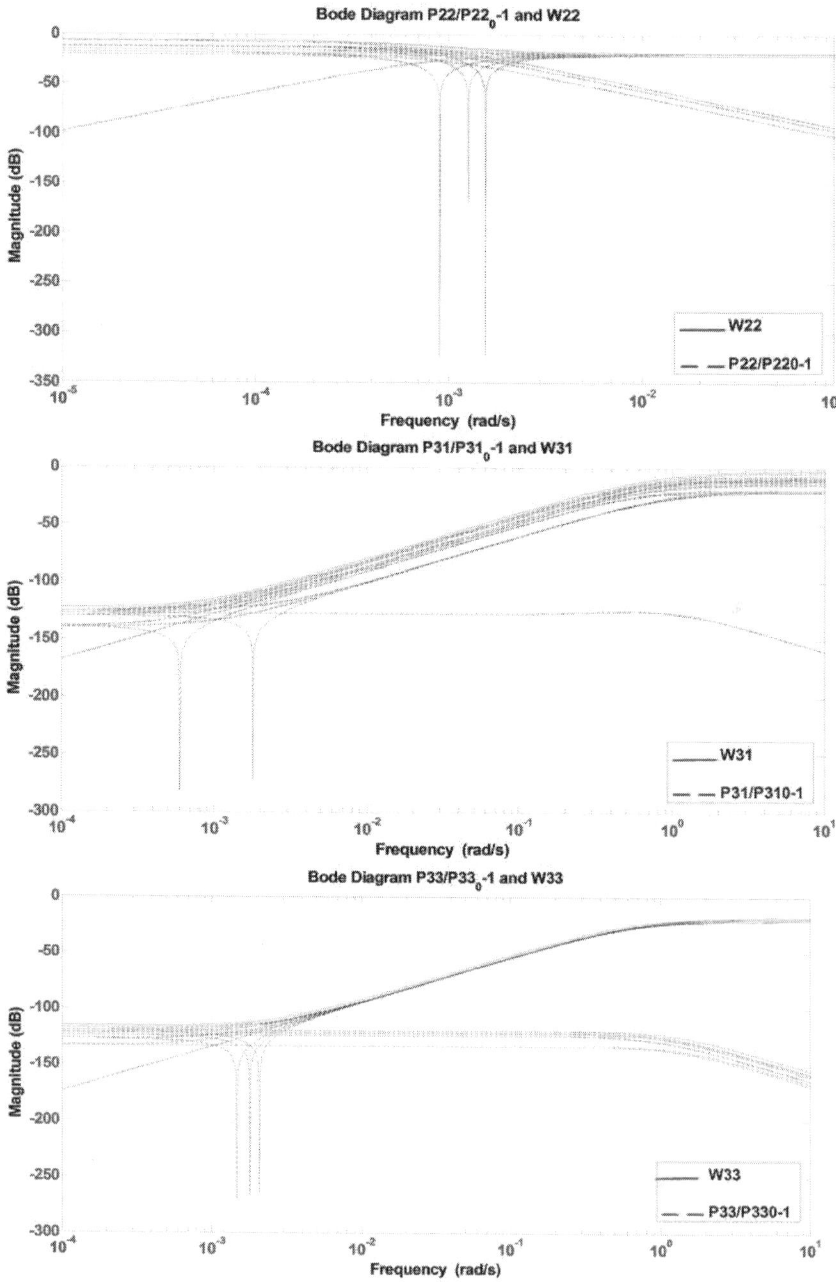

Figure 9. Bode diagram of $\left|P_{ij}(s)/P_{ij_0}(s)-1\right|$ for the non-zero elements of Equation (15).

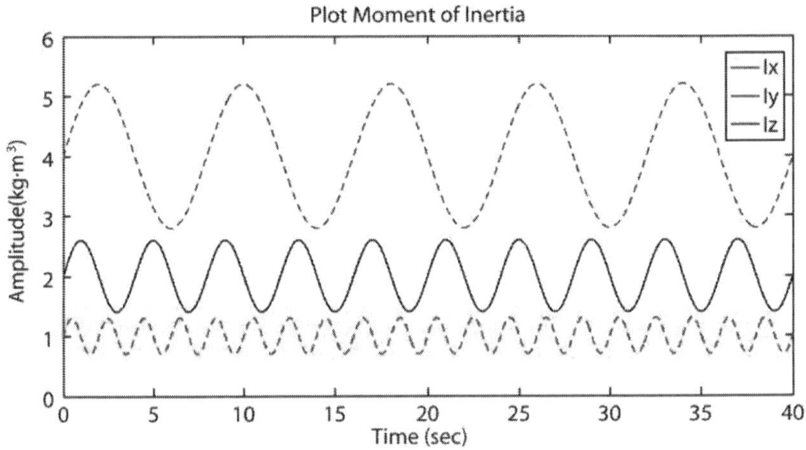

Figure 10. Changes of satellite inertia moments with 30 percentage of uncertainty.

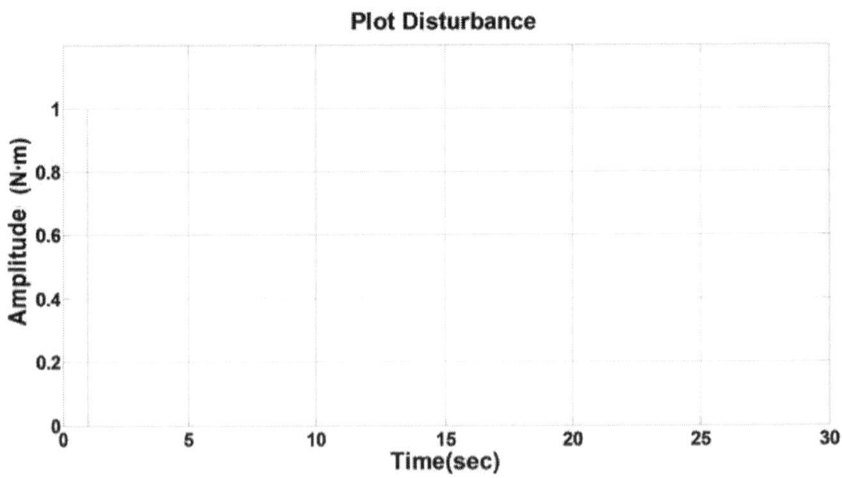

Figure 11. Disturbance pulse input in simulation of the closed-loop system response to disturbance.

Closed-Loop System to Disturbance with $K_{classic}$

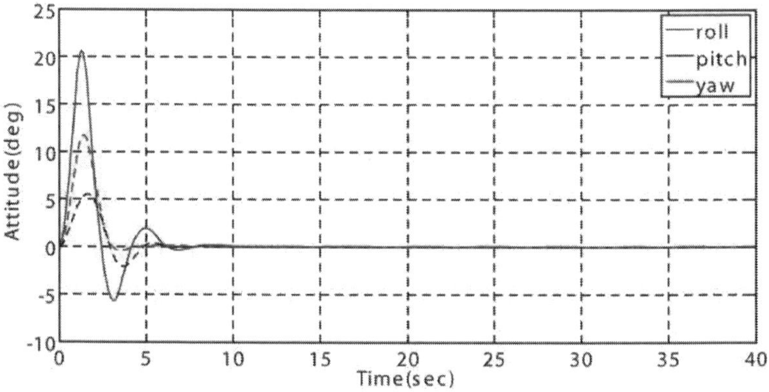

Figure 12. The closed-loop system response with $K_{classic}$ to the disturbance pulse of **Figure 11**.

Closed-Loop System to Tracking with $K_{classic}$

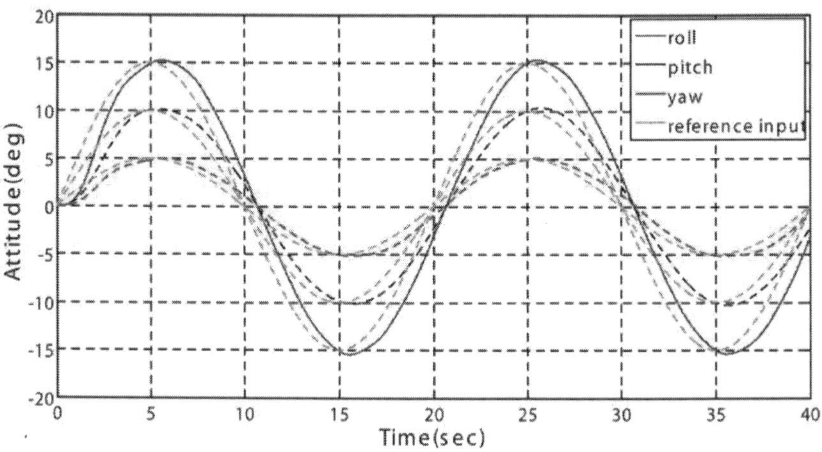

Figure 13. The closed-loop system performance in tracking with $K_{classic}$.

Control Effort with $K_{classic}$

$$\Delta = \left\{ \begin{bmatrix} \Delta_S & 0 \\ 0 & \Delta_F \end{bmatrix} : \Delta_S \in C^{3\times3}, \Delta_F \in C^{6\times6} \right\} \quad (25)$$

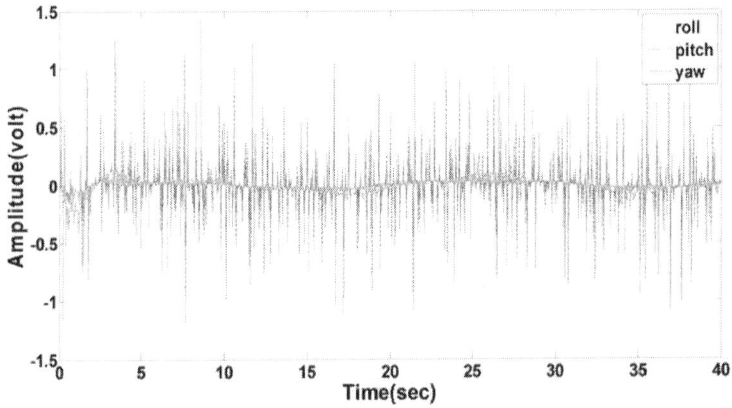

Figure 14. The control effort of closed-loop system with $K_{classic}$ in tracking.

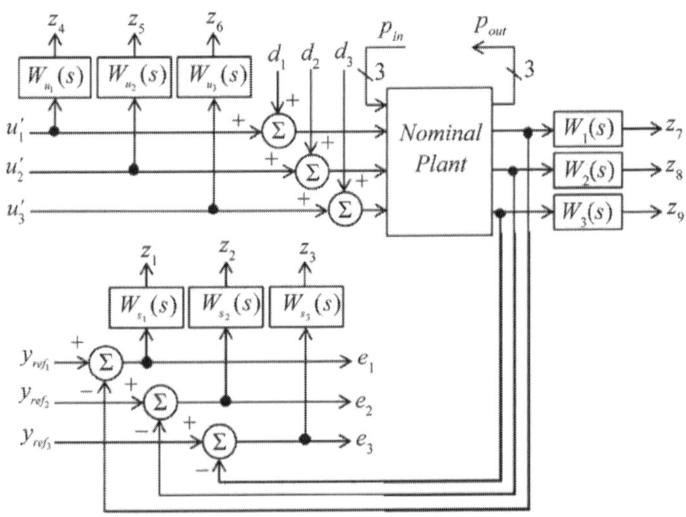

Figure 15. The standard H_∞ configuration.

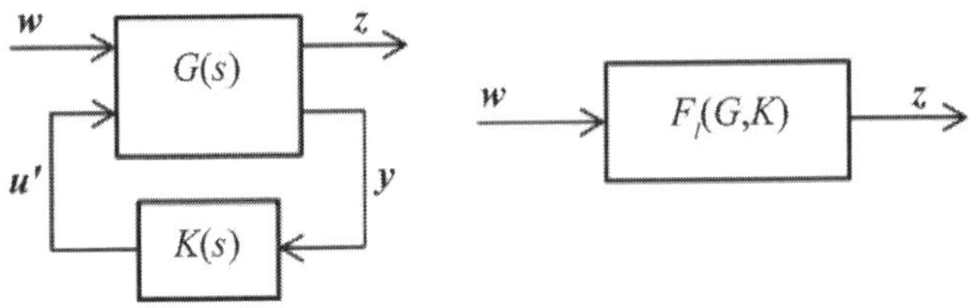

Figure 16. Closed-loop LFT structure in H∞ design.

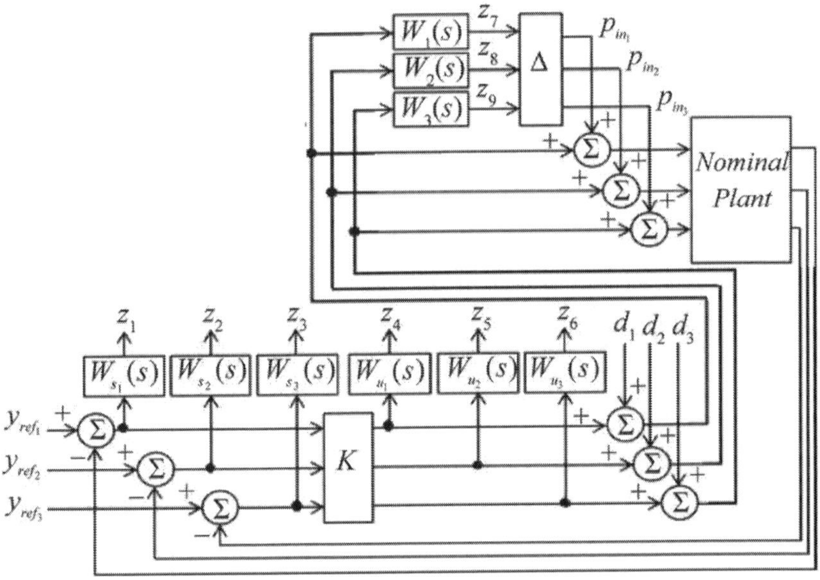

Figure 17. Structure of closed-loop system for μ analysis.

The upper bound and low bound frequency response of μ for H_∞ controller is shown in **Figure 18**. Regarding that for robust stability, the condition

$$\left\| W\left(I+F\left(P_0,\Delta\right)K\right)^{-1} F\left(P_0,\Delta\right)K \right\|_\infty < 1 \tag{26}$$

should be established [14], it could be seen clearly that the closed-loop system with K_{hin} has guaranteed the robust stability.

For nominal and robust performances, the Equations (27) and (28) should be established, respectively [14].

$$\left\| W_s \left(I + F(P_0, \Delta) K \right)^{-1} \right\|_{\infty} < 1 \tag{27}$$

$$\left\| \begin{matrix} W \left(I + F(P_0, \Delta) K \right)^{-1} F(G_0, \Delta) K \\ W_s \left(I + F(P_0, \Delta) K \right)^{-1} \\ W_U \left(I + F(P_0, \Delta) K \right)^{-1} \end{matrix} \right\|_{\infty} < 1 \tag{28}$$

The frequency response of nominal and robust performances is shown in **Figure 19**. It's shown that the system with K_{hin} has the nominal performance but hasn't guaranteed the robust performance.

The closed-loop system's response with K_{hin} to disturbance pulse is drawn in **Figure 20**. As it can be considered, the system in each three channels has eliminated the disturbance with lower overshoot than the classical controller. Its performance in tracking has been also shown in **Figure 21**. In each three channels, the H_{∞} controller has better performance than the classical controller. The control effort in tracking, has been also drawn in **Figure 22**. This control signal is smaller than the corresponding signal for classical controller.

4.2. μ-Synthesis: D-K Iteration Method

The block diagram of the closed-loop system which is used in μ-synthesis is as **Figure 17**. In D-K iteration method, a controller will be obtained from Equation (29).

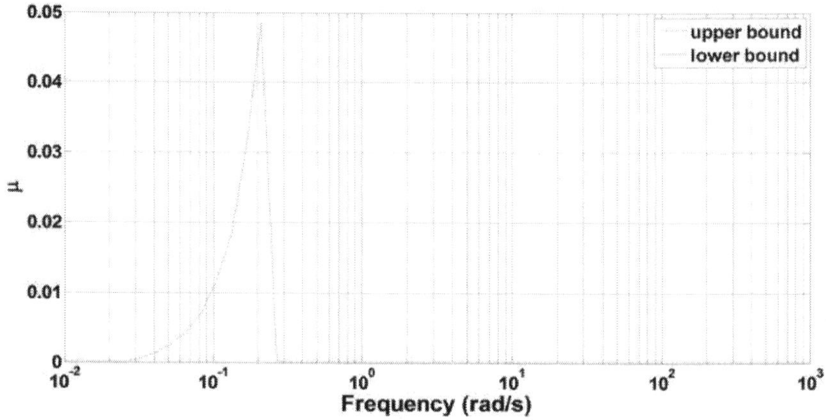

Figure 18. Robust stability of K_{hin}.

Normal Performance and Robust Performance with K_{hin}

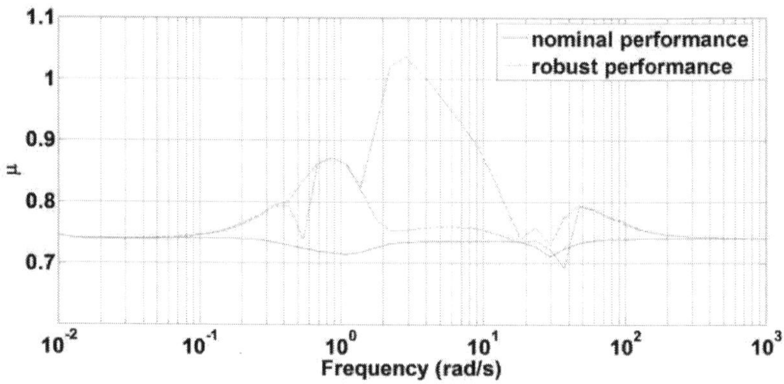

Figure 19. Nominal and robust performance of K_{hin}.

Closed-Loop System Response to Disturbance with K_{hin}

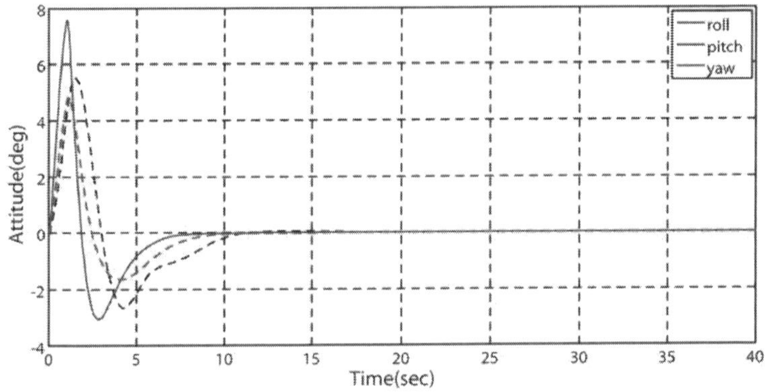

Figure 20. The closed-loop system response with K_{hin} to disturbance pulse in **Figure 11**.

Closed-Loop System Response to Disturbance with K_{hin}

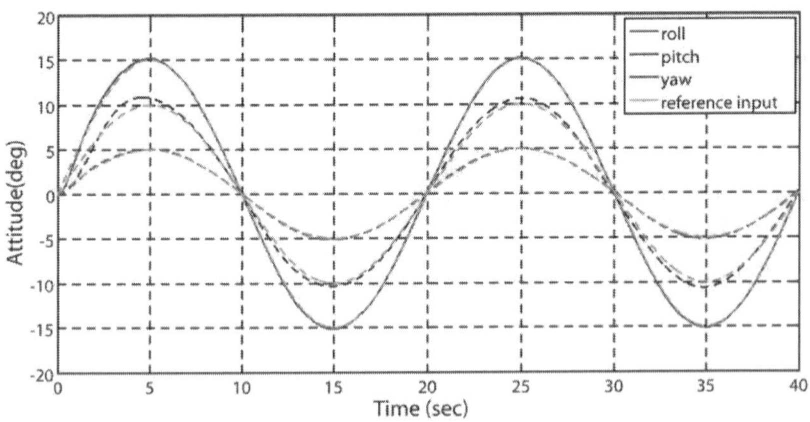

Figure 21. The closed-loop system performance in tracking with K_{hin}.

Control Effort with K_{hin}

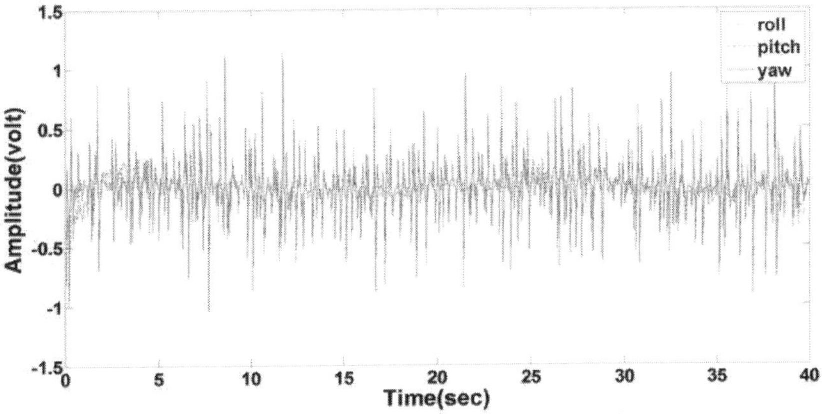

Figure 22. The control effort of closed-loop system with K_{hin} in tracking.

Robust Stability with K_{hin}

$$\min_{\substack{K \\ \text{stabilising}}} \min_{\substack{D_l(s),D_r(s) \\ \text{stabl,} \\ \text{min. phase}}} \left\| D_l(s) F_L(G,K) D_r^{-1}(s) \right\|_\infty \tag{29}$$

The object of μ-synthesis is to find a stabilizing controller of $K(s)$ so that the condition (30), is established [14].

$$\mu_\Delta\left[F_L(G,K)(j\omega)\right] < 1 \tag{30}$$

By using MATLAB [16], after 6 repetitions, the μ - controller has been obtained with $\mu_\Delta = 0.993$. But, the obtained controller is with the degree of 52, which its implementation can be problematic. Usually, a reducedorder controller is preferred. Here for reducing the degree of controller, Hanckel-norm approximation is used [14]. The reduced order controller from the main controller is with the degree of 19. The frequency response of main μ -controller and reduced controller K_{mu} has been shown in **Figure 23**. It can be seen that the frequency response of both controllers are completely on each other.

The robust stability and robust performance analysis of K_{mu} controller are shown in Figures 24 and 25, respectively. It could be considered that contrary to the H_{∞} controller, this controller has guaranteed the robust performance, in addition to robust stability.

The response of the closed-loop system with K_{mu} to disturbance pulse is drawn in **Figure 26**. According to this figure, the system has eliminated the disturbance in each three channels, well. Its performance in tracking and the control effort are shown in Figures 27 and 28, respectively, which are approximately the same as closedloop system with K_{hin}.

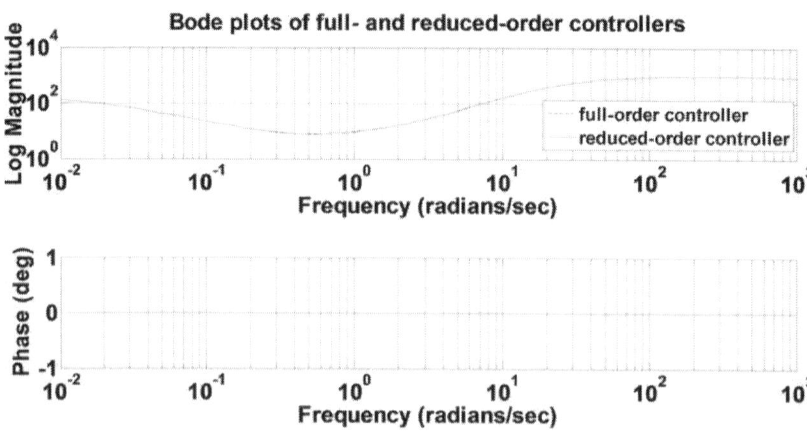

Figure 23. Frequency responses of fulland reduced-order controllers.

Robust stability with K_{mu}

Closed-Loop System Tracking with K_{mu}

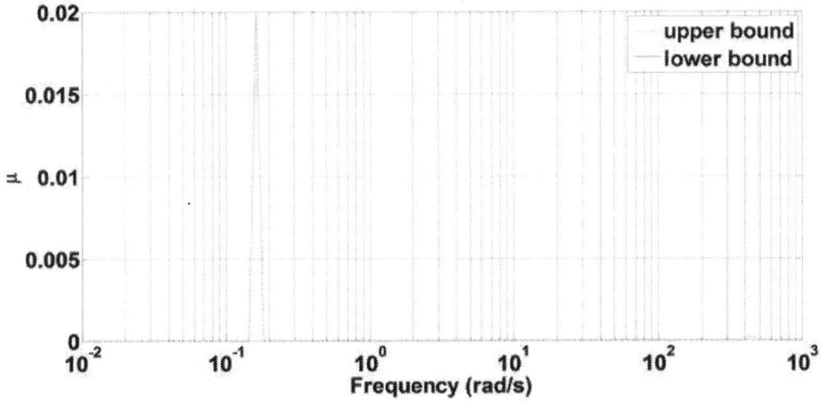

Figure 24. Robust stability of K_{mu}.

Nominal Performance and Robust Performance with K_{mu}

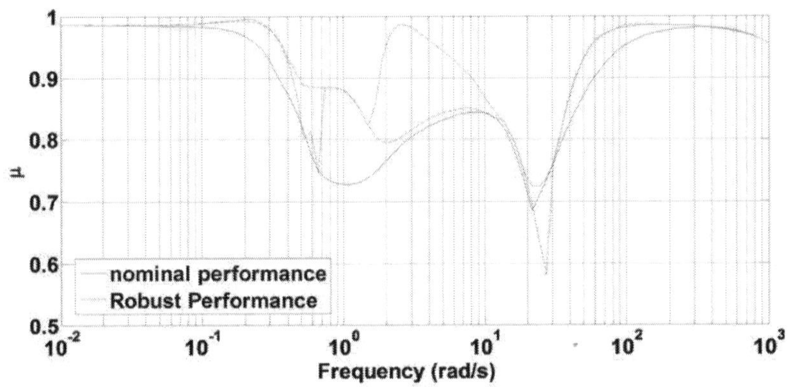

Figure 25. Nominal and robust performance of K_{mu}.

Closed-Loop System Response to Disturbance with K_{mu}

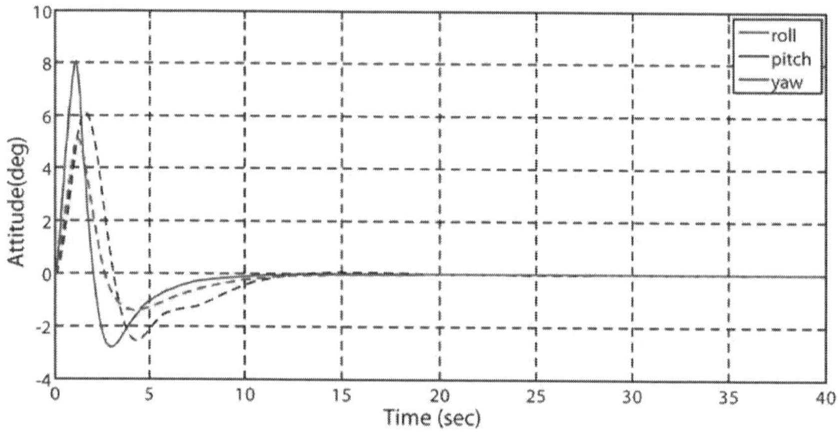

Figure 26. The closed-loop system response with K_{mu} to disturbance pulse in Figure11.

Control Effort with K_{mu}

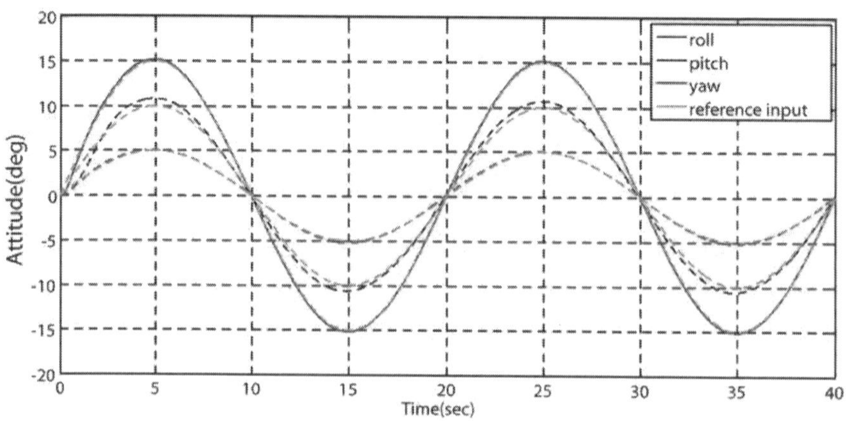

Figure 27. The closed-loop system performance in tracking with K_{mu}.

In **Figure 29**, the performance of two controllers, H_∞ and μ, are compared with increasing of uncertainty norm. This figure shows that μ-controller will guarantee the robust performance for uncertainties with the magnitude of 1.012. For H_∞ controller, the performance of closed-loop will degrade more rapidly with the increase of uncertainty amplitude and won't guarantee the robust performance for uncertainties more than the magnitude 0.9335.

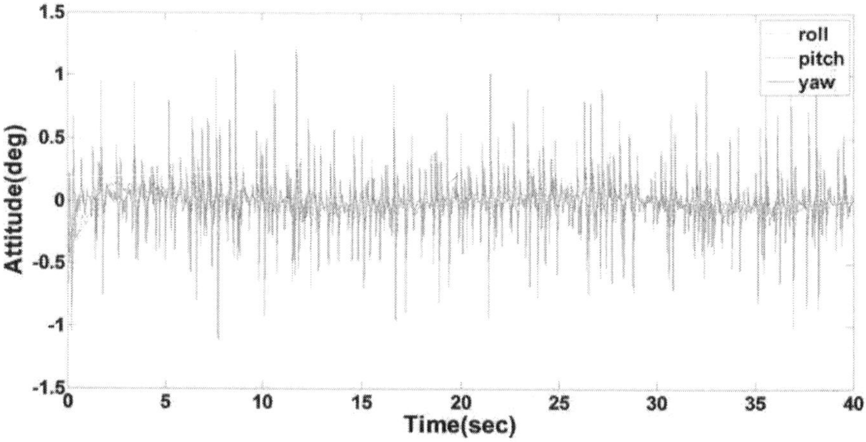

Figure 28. The control effort of closed-loop system with K_{hin} in tracking.

Three designed controllers have been compared with each other in **Table 2**. According to this table, the H_∞ and μ robust controllers have better performance than the classical controller. Although H_∞ and μ controllers performance are approximately the same, but the most important criteria is to guarantee the robust performance. According to **Figure 29** and **Table 2**, only the μ -controller satisfies this criteria; and also the degree of this controller is lower than the H_∞ controller. So, it can be concluded that the μ -controller is the best option for this system comparing to the other controllers.

5. CONCLUSION

The aim of this paper is attitude control of a flexible satellite by using robust controllers. To do so, a classical controller based on Euler degrees' errors and two robust controllers of H_∞ and μ, by using the new idea of internal feedback to resolve robust controller design problem, were designed. The simulation results show that the H_∞ and μ robust controllers are better than classical controller, both in elimination of disturbance and in tracking. The two H_∞ and μ controllers have approximately the same performance regarding the disturbance elimination Performance Degradation Curve for K_{hin} and with K_{mu} and tracking error. But, the μ-controller guarantees the robust performance for the original plant, contrary to the H_∞ controller and has a better performance regarding the robustness against uncertainty of flexible structure of satellite. Furthermore, it has the best performance to satisfy control effort limitations. It has also of lower order in comparison to H_∞ controller. Therefore, it could be concluded that the μ-controller is the best choice among the other controllers.

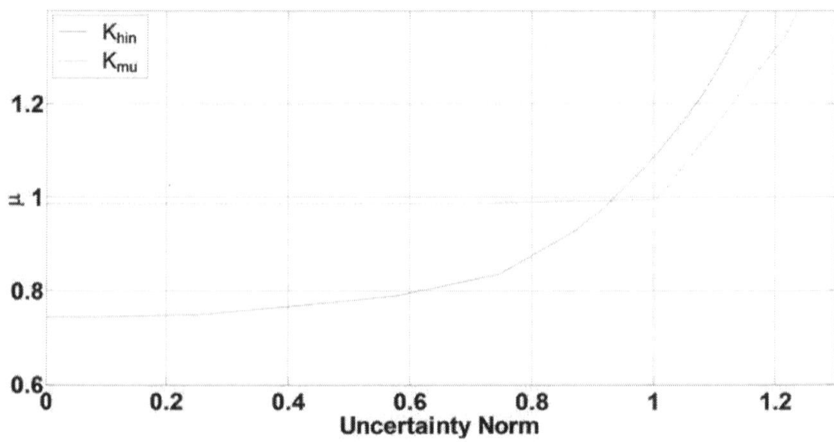

Figure 29. Performance degradation for two controllers.

Table 2. Comparing three designed controllers.

	$K_{classic}$	K_{hin}	K_{mu}
Value of μ in robust performance analysis	-	1.0173	0.987
Overshoot of response to disturbance (deg)	20.5475	7.5880	8.0311
Duration of disturbance elimination with error 0.2 degree (sec)	9.6297	10.3090	11.5372
Error norm in tracking (deg)	5.3089	1.5292	1.5852
Control signal norm in tracking (volt)	35.8124	30.6285	27.8978

REFERENCES

1 D. C. Hyland, J. L. Junkins and R. W. Longman, "Active Control Technology for Large Space Structures," Journal of Guidance, Control and Dynamics, Vol. 16, No. 5, 1993, pp. 801-821.

2 O. Morgul, "Control and Stabilization of a Flexible Beam Attached to a Rigid Body," International Journal of Control, Vol. 51, No. 1, 1990, pp. 11-31.

3 R. B. Noll, J. Zvava and J. J. Deyst, "Effect of Structural Flexibility on Spacecraft Control Systems," Journal of Spacecraft and Rockets, 1969.

4 T. siamak, "On Attitude Recovery of Spacecraft Using Nonlinear Control," Ph.D. Thesis, Concordia University, Montreal, 2005.

5 O. Z. Hayrani, "Dynamics and Control in Modal-Space of Flexible Spacecraft," Ph.D. Thesis, Virginia Polytechnic Institute and State University, Blacksburg, 1979.

6 G.-X. Li, J. Zhou and F.-Q. Zhou, "Variable Structure Control for Flexible Spacecraft," Proceedings of the 25th Chinese Control Conference, Harbin, 7-11 August 2006, pp. 943-946.

7 B. Wu, X. Cao and Z. Li, "Multi-Objective OutputFeedback Control for Microsatellite Attitude Control: An LMI Approach," Acta Astronautica, Vol. 64, No. 11-12, 2009, pp. 1021-1031.

8 M. SeetharamaBhat, A. G. Sreenatha and S. K. Shrivastava, "Robust Low Order Dynamic Controller for Flexible Spacecraft," IEE Proceedings-D, Vol. 138, No. 5, 1991, pp. 460-468.

9 C. D. Yang and Y. P. Sun, "Mixed H_2/H_∞ State-Feedback Design for Microsatellite Attitude Control," Control Engineering Practice, Vol. 10, No. 9, 2002, pp. 951-970.

10 J. S. Chae and T. W. Park, "Dynamic Modeling and Control of Flexible Space Structures," KSME International Journal, Vol. 17, No. 12, 2003, pp. 1912-1921.

11 W. Wang, P. P. Menon, D. G. Bates and S. Bennani, "Robustness Analysis of Attitude and Orbit Control Systems for Flexible Satellites," IET Control Theory & Applications, Vol. 4, No. 12, 2010, pp. 2958-2970.

12 M. J. Sidi, "Spacecraft Dynamics and Control a Practical Engineering Approach," Cambridge University Press, New York, 1997.

13 J. Kulkarni and M. Campbell, "An Approach to Magnetic Torque Attitude Control of Satellites via H_∞ Control for LTV Systems," 43rd IEEE Conference on Decision and Control, Bahamas, 14-17 December 2004, pp. 273-277.

14 S. Skogestad and I. Postlethwaite, "Multivariable Feedback Control Analysis and Design," John Wiley & Sons, Hoboken, 2001.

15 K. Zuou and J. Doyle, "Essentials of Robust Control," Prentice Hall, Upper Saddle River, 1998.

16 D.-W. Gu, P. Hr. Petkov and M. M. Konstantinov, "Robust Control Design with Matlab," Springer, Berlin, 2005.

CHAPTER 2

Robust Flight Control Design to Minimize Aircraft Loss-of-Control Incidents

Ronald A. Hess

Department of Mechanical and Aerospace Engineering, University of California, One Shields Ave., Davis, CA 95616, USA; Tel: 530-752-1513

ABSTRACT

A pseudo-sliding mode control synthesis procedure discussed previously in the literature is applied to the design of a control system for a nonlinear model of the NASA Langley Generic Transport Model. The complete vehicle model is included as an appendix. The goal of the design effort is the synthesis of a robust control system to minimize aircraft loss-of-control by preserving fundamental pilot input—system response characteristics across the flight envelope, here including the possibility of actuator damage. The design is carried out completely in the frequency domain and is described by a ten-step synthesis procedure, also previously introduced it the literature. Five different flight tasks are considered in computer simulations of the completed design demonstrating the stability and performance robustness of the control system.

KEYWORDS

Flow and flight control systems; guidance; navigation and control

1. INTRODUCTION

Transport aircraft loss-of-control (LOC) incidents are a continuing problem in aviation safety [1,2]. There is growing interest in the design and validation of robust flight control systems to alleviate the severity of LOC incidents, e.g., [3,4]. A majority of these control designs are adaptive in nature. It is of interest to explore the feasibility of employing non-adaptive systems in this role. The research described herein employs frequency-domain sliding mode control to provide a robust flight control system for a model of the NASA Langley Generic Transport Model (GTM) [5], shown in Figure 1. Although referred to as a "model" the vehicle is actually a 5% dynamically-scaled version of a commercial transport. The frequency-domain sliding mode control approach has been applied in a number of instances in the literature, e.g., [6,7,8,9,10,11,12,13]. The contribution of the present study lies in the application of the frequency-domain sliding mode control to an existing flight vehicle (GTM) which has been the subject of other studies, e.g., [14] and in which a set of nonlinear dynamics have been explicitly and succinctly described [15]. The philosophy behind the approach espoused herein is that the preservation of fundamental pilot input—system response characteristics throughout the flight envelope is important, indeed vital, in preventing LOC events. Obviously, this is not a novel viewpoint. Indeed, it is shared by researchers developing model reference adaptive control designs, e.g., [14]. Among other potential outcomes, sudden changes in aircraft response characteristics can create "triggering events" leading to pilot-induced oscillations (PIOs) [16]. Quoting findings and recommendations from [16]:

"**Finding 1-1** Adverse (PIO) events are fundamentally interactive and occur during highly demanding tasks when environmental, pilot, or *aircraft dynamic changes create or trigger mismatches between actual and expected aircraft responses* (emphasis added)."

"**Recommendation 4-3** Organizations should adopt and implement minimization techniques in *design* and development policies, *processes*, and procedures. These techniques should be tailored and routinely updated to accommodate applications of newly developed technologies (emphasis added)."

Figure 1. The NASA generic transport model.

The paper is organized as follows: Section 2 and the Appendix describe the nonlinear GTM model. Section 3 outlines the frequency-domain sliding mode design procedure, and then applies the procedure to the design of a flight control system for the GTM vehicle. The reader is reminded that the introductory material of Section 3 has appeared before in the literature, e.g., [13]. Section 4 discusses the design of the outer-loop compensation elements. Section 5 provides computer simulations of five tasks for the GTM vehicle. Section 6 provides a discussion of the results. Section 7 provides conclusions to the study.

2. VEHICLE MODEL

The appendix provides a nonlinear model of the vehicle in question, the NASA Langley GTM. These dynamics are taken directly from [15] and repeated here for the convenience of the reader. The study of [10] was devoted to bifurcation analyses of the nonlinear model, with applications to the study of LOC events. The Appendix also includes elevator, aileron and rudder actuator models (including amplitude and rate limits) obtained from [17]. The inclusion of these models allows the reader to (1) replicate the results of the author and/or (2) compare the results with competitive designs.

3. FREQUENCY DOMAIN SLIDING MODEL CONTROL SYSTEM DESIGN

3.1. The Design Procedure

- The design procedure for the frequency-domain sliding mode control approach has been thoroughly described in the literature, e.g., [13]. For the sake of completeness, it will be repeated here:
- A vehicle model is obtained. If the vehicle in question is nonlinear, a linearized version is obtained for initiating the SMC design. Actuator dynamics are ignored at this stage. The linearization here is only evoked to allow classical frequency-domain design techniques to be employed in the synthesis procedure to be described.
- A square control structure is identified. The design requires that no transmission zeros lie in the right-half plane and any uncontrollable states must be asymptotically stable. These requirements are similar to those employed in feedback linearization designs. Observability of the linear model must be in evidence to accommodate the observer design to be discussed in Step (7). The control loop structure in the square system is based upon a simple premise. Consider a square structure containing "m" control loops eventually driving a single control variable u_c. Define the inner-most loop as loop "1", with the next being "m − 1", *etc*. The transfer function of the linearized vehicle model for the control loop "1" should have a relative order of unity. The command variable in the "m − n" loop is, over a limited but important frequency range, the approximate integral of the command variable in the "m − (n + 1)" loop. Sequential loop

closure design like that just described is certainly not novel, e.g., [18]. Its use here, however, forms the basis of the step-by-step design procedure being described.

Rules (3)–(10) refer to the inner-most control loops in the square structure.

Sliding mode control is limited to the inner-most control loops in the structure. The control law for each of these loops is given by

$$u_c(t) = \rho \operatorname{sgn}(\sigma) \tag{1}$$

and the sliding manifolds σ for each control channel are selected.

a) σ for any channel is derived from a tracking error expression such as

$$\sigma = e(t)^{\xi-1} + K_{\xi-2}e(t)^{\xi-2} + \cdots + K_0 e(t) + K_{-1}\int e(t)dt \tag{2}$$

where ξ is the relative order of the system. Here $e(t)^{\xi-i}$ refers the $(\xi-i)^{th}$ derivative of e(t). An integral term also appears in Equation (2) to counter the steady-state bias often created with the inclusion of a boundary layer.

b) Recognizing that a boundary layer is to be implemented, the control law is expressed as a linear transfer function,

$$u_c(s) = \rho\sigma(s) = \rho\left\{\left[s^{\xi-1} + K_{\xi-2}s^{\xi-2} + \cdots K_0 + \frac{K_{-1}}{s}\right]\cdot e(s)\right\} \tag{3}$$

The parameters K_i are chosen to provide "desirable" properties in the frequency domain. This means creating a loop transmission with broad K/s-like characteristics around crossover. The minimum K_ρ is that which permits pure gain or PI compensation in the frequency range at and beyond which $\xi = 1$, while exhibiting adequate gain and phase margins.

With the $K_{(\cdot)}$ values just determined, the existence of sliding behavior is verified through computer simulation. A convenient and practical choice for ρ is the amplitude limit of the actuator for the loop in question. Thus $u_c(t)$ is the output of a relay element with limits of $\pm\rho$.

- A boundary layer is included in the controller to eliminate the high-frequency switching in the control variable(s) $u_c(t)$.

- Actuator dynamics are now included in the computer simulation. Instability will typically result. This condition highlights the effect of so-called "parasitic dynamics" in the SMC design procedure.

- Asymptotic observers are created for each channel with actuator dynamics still eliminated in the observer design. The eigenvalues of the observers are selected as real and (approximately) equal, and are determined by either maximizing the stability margins evident in "effective" unity feedback loop transmissions L_{eq}, e.g., [13], or simply by selecting the eigenvalues near the bandwidth of the actuators.

- To increase robustness, "hedge" dynamics can be created in which an additional loop is closed in parallel with the observer. Briefly, the hedge dynamics further decrease the destabilizing effects of the actuator in the loop in question. The method for choosing hedge dynamics can be found in [10] and will be summarized here. Hedge dynamics are created from a Bode diagram of the hedge dynamics transfer function G_{hedge} as follows: The magnitude plot of the hedge dynamics should exhibit (a) a +20 dB/dec slope at low frequencies, (b) a -20ξ dB/dec slope at frequencies near the actuator bandwidth, where ξ is the relative order of the vehicle dynamics in the loop in question (excluding actuators), and (c) a $(-20\xi - 20)$ dB/dec slope at high frequencies. The gain value is chosen so that the magnitude of the Bode diagram of the hedge dynamics equals that of the transfer function of the appropriate observer output to SMC output near the natural frequency of the appropriate actuator [10].

In some applications, the addition of a rate-feedback loop in the inner-most control loops to which FDSMC is being applied can be considered. This was done, for example in [19]. Here "rate" refers to the fact that the derivative of the output of the inner-most feedback loop is utilized. In this case, it would be pitch and roll acceleration.

If desired, the observer(s) of Step (7) can be scheduled with flight condition.

3.2. Design Specifics for the GTM Vehicle

The vehicle model in the Appendix was mechanized in Simulink® and linearized about a flight condition of steady-wings-level flight at a trim velocity of 110 ft/s. This step required an estimation of trim values of pitch attitude, θ_0, angle of attack, α_0, thrust level, T_0, and elevator angle, δe_0. These were estimated as: $\theta_0 = 5/57.3$ rad, $\alpha_0 = 5/57.3$ rad, $T_0 = 5$ lbf, $\delta e_0 = 0$ rad. The results of [15] provided guidance in choosing these values. It should be emphasized that after the SMC design is complete, the initial trim value can be obtained exactly, as the final SMC controller will "self-trim" in simulation. For reference, the following pitch and roll-rate open-loop transfer functions were obtained at this stage:

$$\frac{q}{\delta_e} = \frac{-13.1736s(s+2.166)(s+.06364)}{(s^2+0.0129s+0.1443)(s^2+5.716s+35.34)} \tag{4}$$

$$\frac{p}{\delta_a} = \frac{-10.8256(s-.02506)(s^2+1.929s+23.5)}{(s+5.207)(s+0.0597)(s^2+1.758s+28.02)} \tag{5}$$

The reader will recognize typical modal characteristics in Equations (4) and (5), *i.e.*, short period, phugoid, roll subsidence, spiral, and Dutch-roll modes [20]. Also note the non-minimum phase zero in Equation (5), not uncommon in such transfer functions [21].

Figure 2 shows the Bode diagrams of the transfer functions of Equations (4) and (5), indicating relative orders of 1 (−20 dB/dec magnitude slopes beyond 10 rad/s).

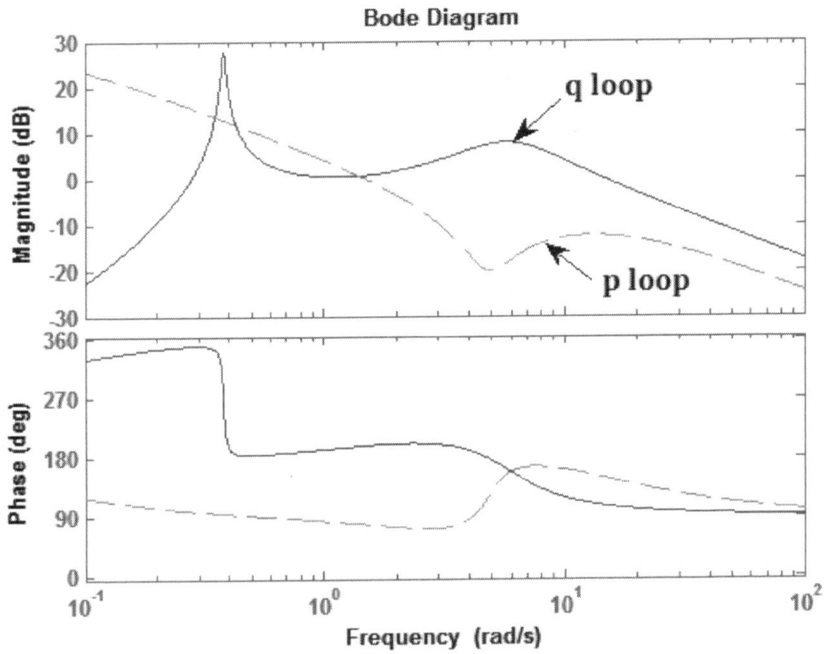

Figure 2. Bode Diagrams for the Transfer Functions of Equations (4) and (5).

The control structure to be used is shown in simplified form in Figure 3. Note that no yaw control system is to be created. In terms of pilot control of a full-scale aircraft, the system of Figure 3 would define flight-path control with the cockpit column, roll-rate control with the wheel, auto-throttles controlling airspeed, with bare-airframe dynamics in response to pedal inputs.

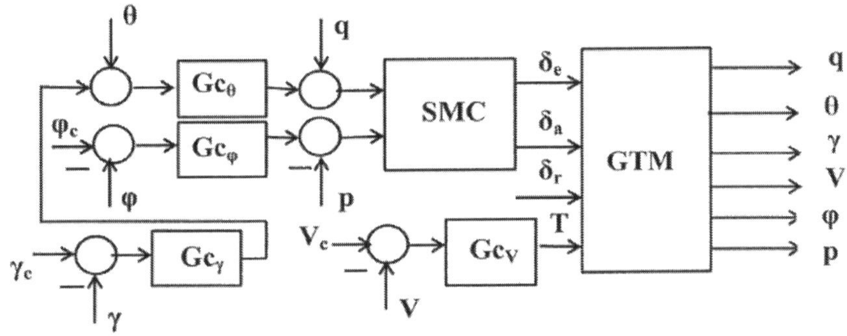

Figure 3. The control system architecture.

The sliding manifolds for the q and p loops, as interpreted in the frequency domain can be given by

$$Gc_q = \frac{-0.658(s+7)}{s} \tag{6}$$

$$Gc_p = \frac{-1.8(s+6)}{s} \tag{7}$$

The simplicity of these compensators should be noted. To accommodate the non-minimum phase characteristics of Equation (5), "regulated variables" were employed for the feedback variable in the p loop p(t) + 0.5r(t). Regulated variables were also used in the study of [22]. The gains shown in Equations (6) and (7) result in crossover frequencies of 20 rad/s. The author would point out that this value is reasonable for a 5% scaled aircraft.

By introducing initial conditions of 5/57.3 rad/s in the q-loop and −5/57.3 rad/s in the p-loop sliding behavior was verified. That is, "reaching" and "sliding" behavior was noted along with "infinite frequency" switching in the outputs of the signum elements. Figure 4 shows the reaching and sliding behavior in the p loop. Note that for sliding behavior, the numerator of Equation (7) is zero, *i.e.*, p(t) = −6φ(t) on the sliding surface [23].

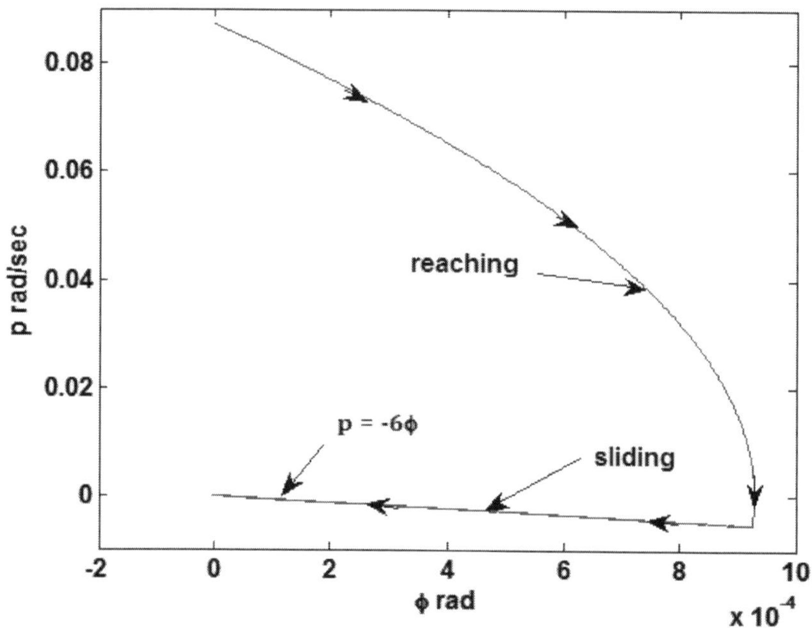

Figure 4. Reaching and sliding behavior in the p loop.

A boundary layer was included in the q and p loops to eliminate the high-frequency switching characteristic of SMC systems. The signum elements in the q and p loops were removed. As in other applications of the methodology, e.g., [13], no saturation elements were included. This leads to an accurate description of the methodology as *pseudo-sliding* mode control.

Actuator dynamics were next included. Due to the relative high-frequency bandwidth of the actuators, instability was avoided in this case. Nonetheless, asymptotic observers were created for the q and p loops to ensure robustness in the presence of the actuator dynamics. A single observer was utilized with inputs consisting of φ, θ, ψ, p, q, r and the outputs of all the inner-loop compensators. The eigenvalues were selected as $\lambda = -75, -75.1, -75.2, \ldots, -75.8$, just beyond the actuator bandwidths of 62.8 rad/s.

Hedge dynamics were created in the q and p loops as follows:

$$q\ loop: \frac{-s}{0.006667s^3 + 1.267s^2 + 42.67s + 400} u_{cq} \tag{8}$$

$$p\ loop: \frac{-3.63s}{0.006667s^3 + 1.267s^2 + 42.67s + 400} u_{cp} \tag{9}$$

where u_{cq} and u_{cp} are the output of the elements in Equations (6) and (7).

Rate feedback loops were included in the q and p loops as follows:

$$q\ loop: \frac{-s}{0.006667s^3 + 1.267s^2 + 42.67s + 400} u_{cq} \tag{10}$$

$$p\ loop: \frac{0.05s}{[.0025s + 1]}(q + 0.05r) \tag{11}$$

Figure 5 shows the manner in which the hedge and rate-feedback dynamics were employed, in this case, for the q loop. As an example, Figure 6 shows the result of adding the rate feedback loop to the loop transmission of the p loop. The phase difference in Figure 6 merely represents a 360 deg shift. A significant reduction in phase lag at high frequency can be seen.

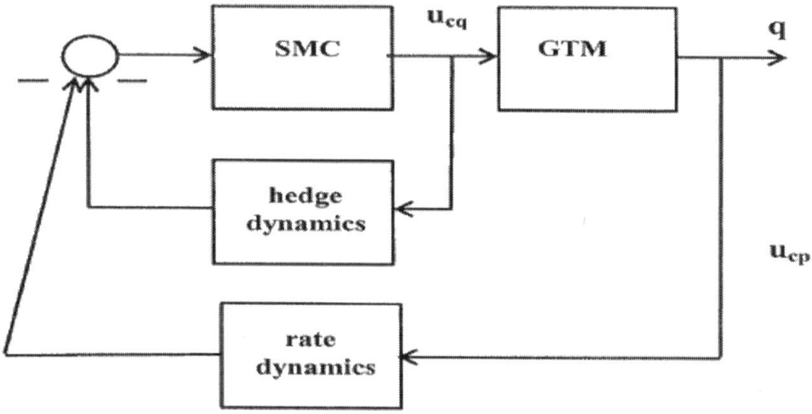

Figure 5. Hedge and rate dynamics placement.

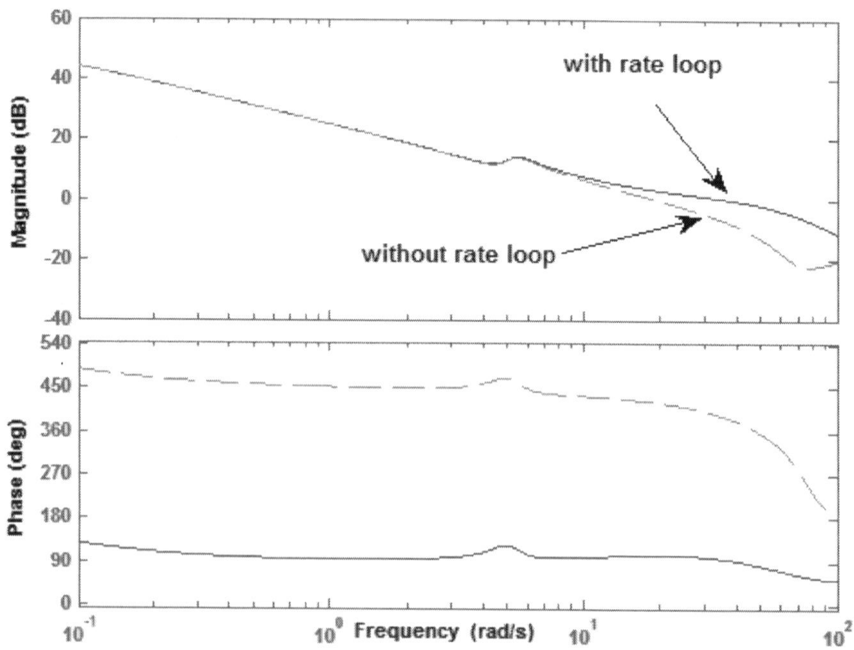

Figure 6. Effect of rate-loop closure in loop transmission of p loop.

For the purposes of this study, two flight conditions were considered, both at sea level, one at 110 ft/s and one at 220 ft/s. The asymptotic observers were scheduled with respect to these flight conditions.

4. DESIGN OF OUTER-LOOP COMPENSATORS

The four non-SMC compensators in Figure 3 were created using classical loop-shaping principles, *i.e.*, ensuring K/s-like characteristics in the loop transmissions and separating the crossover frequencies of each sequential loop by at least a factor of 3. The resulting compensation elements are given in Equation (12).

$$Gc_\theta = 7.16,\ Gc_\gamma = \frac{1.08(s+2)}{s},\ Gc_\varphi = \frac{7.16(s+2)}{s},\ Gc_V = \frac{5.26(s+0.5)}{s} \qquad (12)$$

The simplicity of these elements is worthy of note as each is either a pure gain or PI controller.

5. COMPUTER SIMULATION OF THE DESIGN

5.1. Sensor Noise and Unmodeled Time Delays

Additive sensor noise was included in each of the response variables used in the observers. The noise was created by passing white noise second-order filters with break frequencies of 10 rad/s. The root-mean square noise levels were 0.1 deg, 0.1 deg/s on vehicle attitudes and attitude rates, respectively. In addition, unmodeled time delays of 0.2 s were introduced at the input of each actuator.

5.2. Task 1 and Actuator Damage

A computer simulation of the nonlinear vehicle and control system was undertaken with the following task:

1) Command flight path angle $\gamma = 0$ deg.

2) Command velocity $V_c = 110$ ft/s

3) Command velocity perturbations around trim $\Delta V_c = 15\ \sin(0.1\ t)$ ft/s

4) Command roll attitude $\varphi_c = 10\ \sin(t)$ deg and simultaneous rudder input $\delta_r = 5\ \sin(t)$ deg. The roll and rudder commands were deliberately designed to produce positive (negative) sideslip β, with positive (negative) roll attitude φ. These inputs were included to excite the nonlinearities in the vehicle model. While decidedly atypical pilot inputs, they have been seen in piloted simulations involving crosswind landings [24].

To emulate damage to the elevator and aileron actuators, at 30 s into a 180 s run the effectiveness of the actuators was reduced by 25% and a 0.5 deg backlash element was introduced in the outer loop of the actuator mechanization. Simultaneously a −10 deg elevator bias (trailing edge up) was introduced.

Figure 7 and Figure 8 show the vehicle responses and longitudinal control/thrust inputs. Note that the angle of attack reaches approximately 10 deg in Figure 7. Figure 8 indicates that the elevator bias was immediately compensated by the SMC system. By way of comparison, Figure 9 and Figure 10 show the vehicle responses and longitudinal control/thrust inputs when no actuator damage was considered. It is interesting to note that the vehicle responses in the two cases are essentially identical.

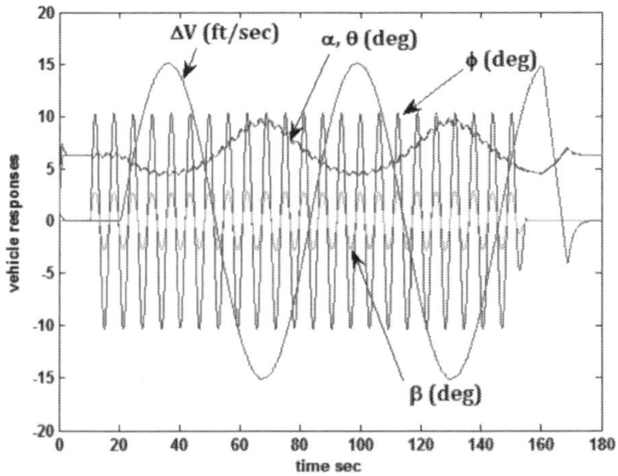

Figure 7. Vehicle responses for Task 1.

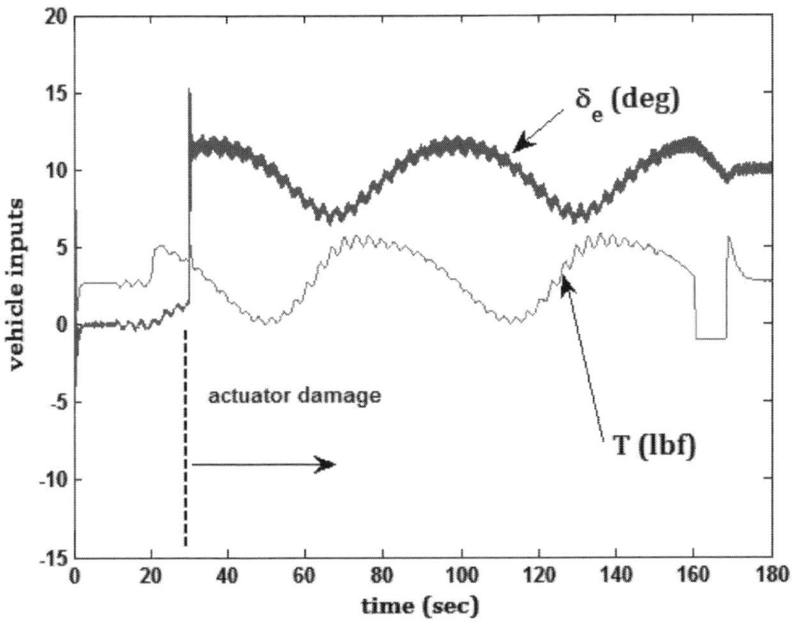

Figure 8. Elevator and thrust inputs for Task 1.

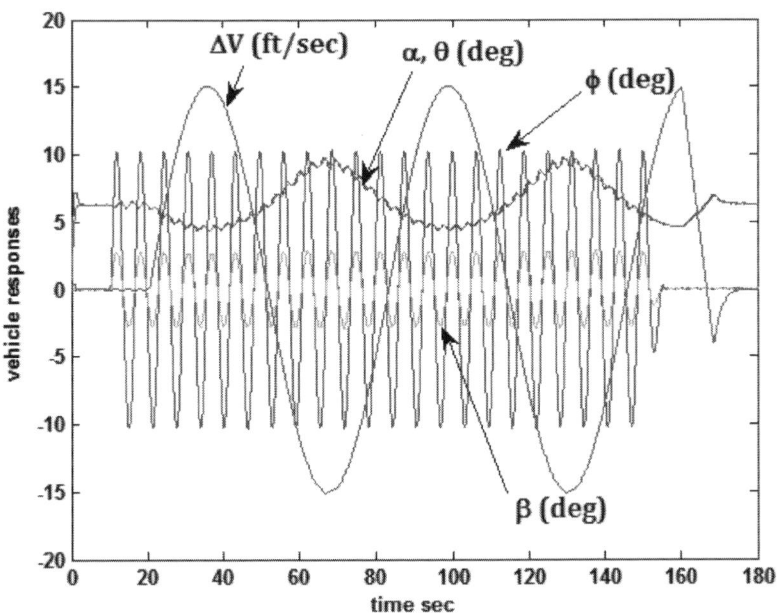

Figure 9. Vehicle responses for Task 1 with no actuator damage.

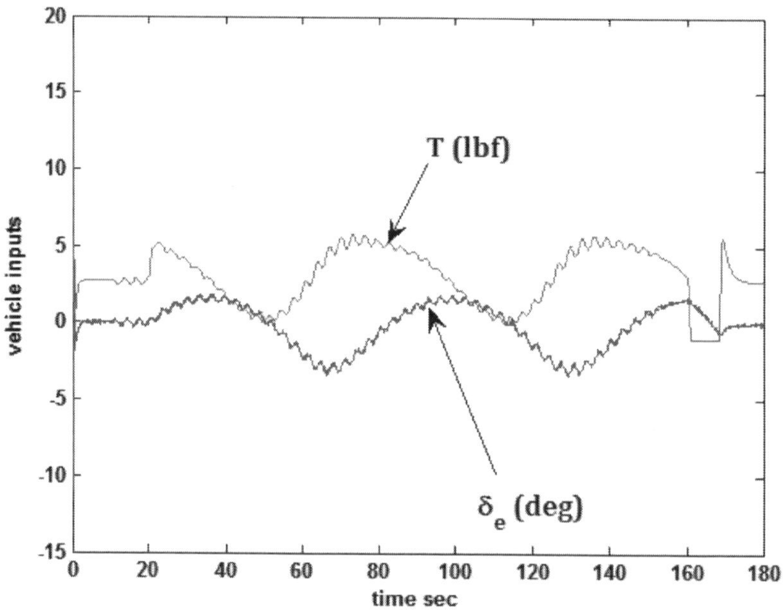

Figure 10. Elevator and thrust inputs for Task 1 with no actuator damage.

5.3. Task 2 and Actuator Damage

The scenario of Task 1 was repeated, with the command velocity V_c = 220 ft/s. Figure 11 and Figure 12 show the results. Note in Figure 13 that $\delta_e - 10$ (deg) has been plotted to prevent overlap with the thrust time history. The very active elevator inputs are attributable to the effect of the actuator backlash element at the higher airspeed.

5.4. Task 3 and Actuator Damage

Task 3 consisted of the following

1) Command flight path angle $\gamma = 2.5$ deg.

2) Command velocity $V_c = 110$ ft/s

3) Command velocity perturbation around trim $\Delta V_c = -22.5$ ft/s

4) Command roll attitude $\varphi_c = 10 \sin(t)$ deg and simultaneous rudder input $\delta_r = 5 \sin(t)$ deg.

The actuator damage described in Section 6 was also included. Figure 13 and Figure 14 show the results. Note in Figure 14that the angle of attack is approaching 13 deg, quite near stall for this vehicle [25].

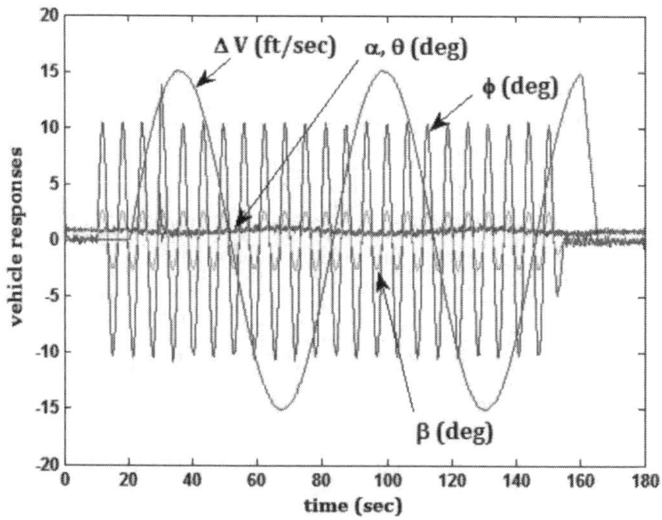

Figure 11. Vehicle responses for Task 2.

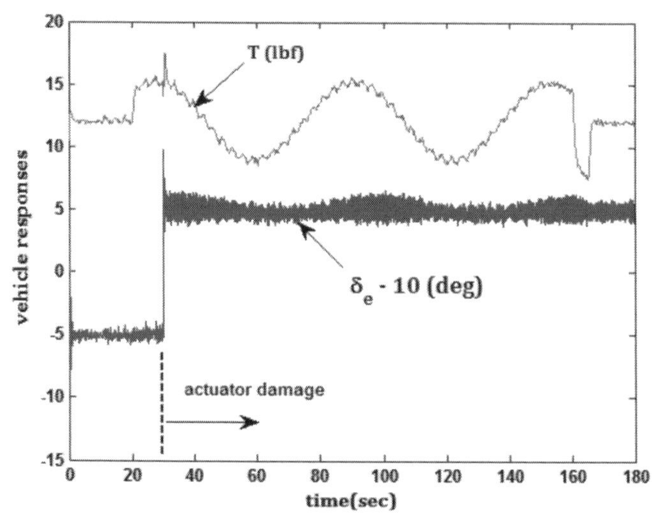

Figure 12. Elevator and thrust inputs for Task 2.

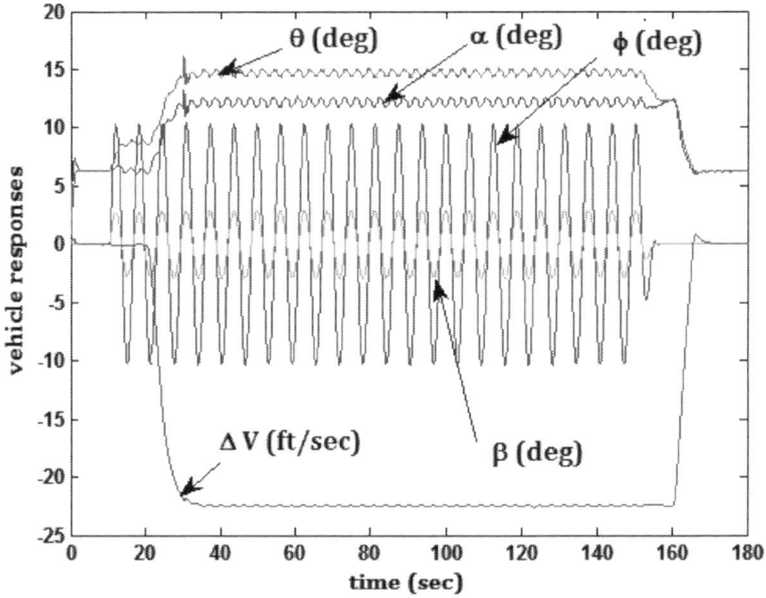

Figure 13. Vehicle responses for Task 3.

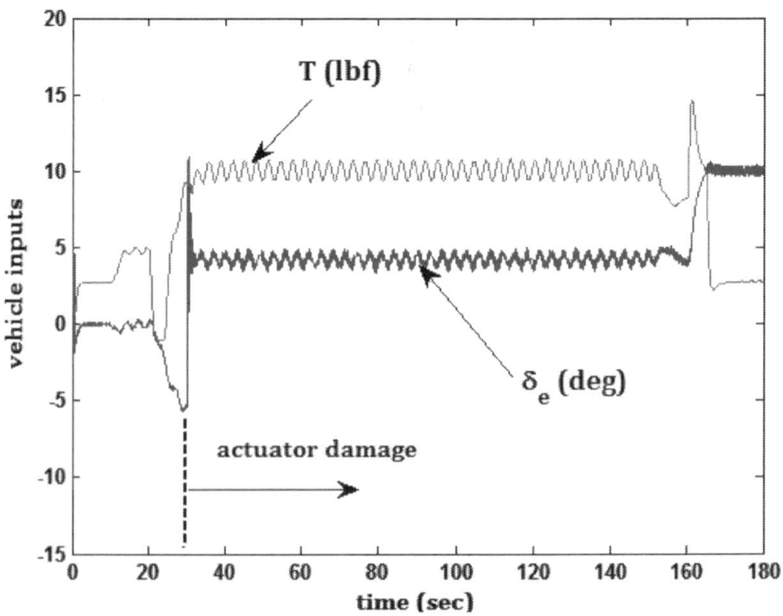

Figure 14. Elevator and thrust inputs for Task 3.

5.5. Task 4 and Actuator Damage

1) Task 4 was identical to Task 3 with the following two exceptions:

2) Command velocity perturbation around trim $\Delta V_c = -7.5$ ft/s

Air density in the vehicle model was set to a value appropriate for an altitude of 10,000 ft ($\rho = 0.0017556$ slugs/ft^3). The author is aware that the vehicle in question was not intended to operate at this altitude. The inclusion of this flight condition, however, provides an opportunity to evaluate the robustness of the SMC design.

The necessity of reducing the magnitude ΔV_c from that of Task 3 was necessary to prevent stall at the higher simulated altitude. Figure 15 and Figure 16 show the vehicle responses and inputs. The maximum angle of attack in the task was approximately 12 deg.

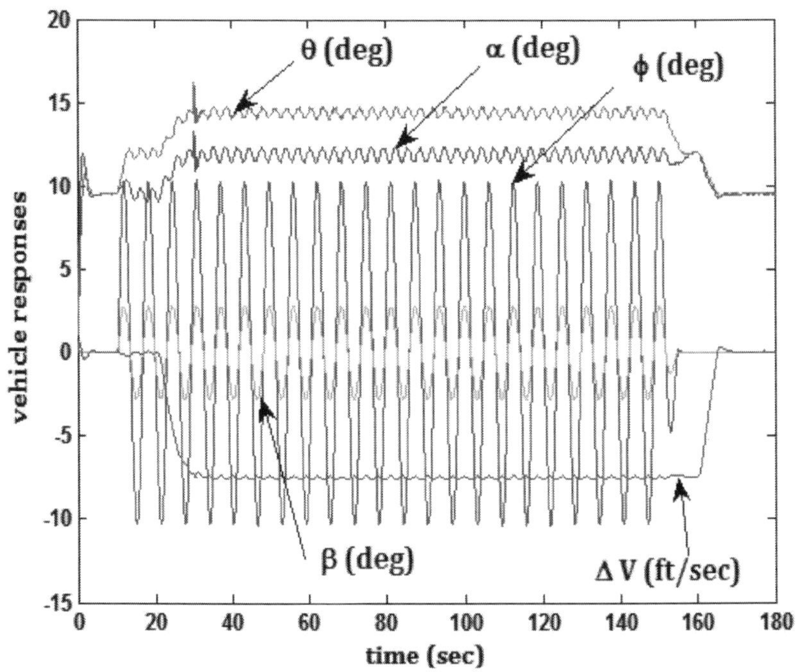

Figure 15. Vehicle responses for Task 4.

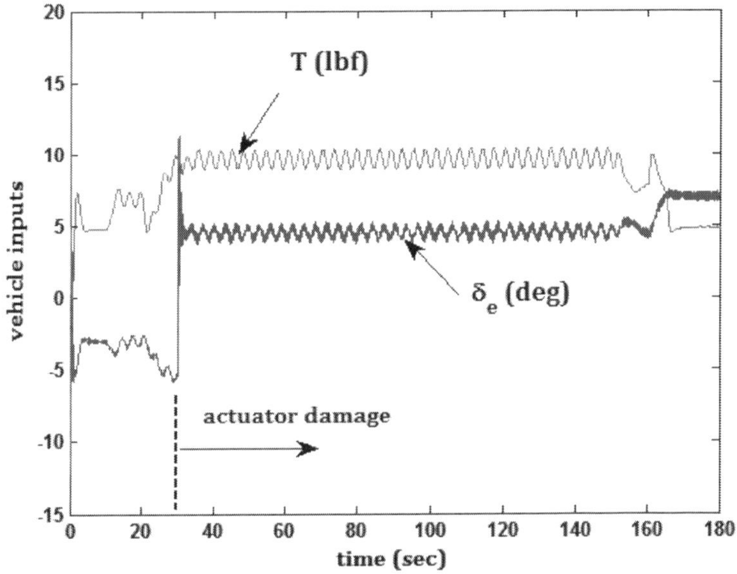

Figure 16. Elevator and thrust inputs for Task 4.

5.6. Task 5 and Actuator Damage

Task 5 was identical to Task 4 save that the trim airspeed was 220 ft/s. Figure 17 and Figure 18 show the simulation results. Again, the elevator activity is attributable to the higher airspeed and actuator damage.

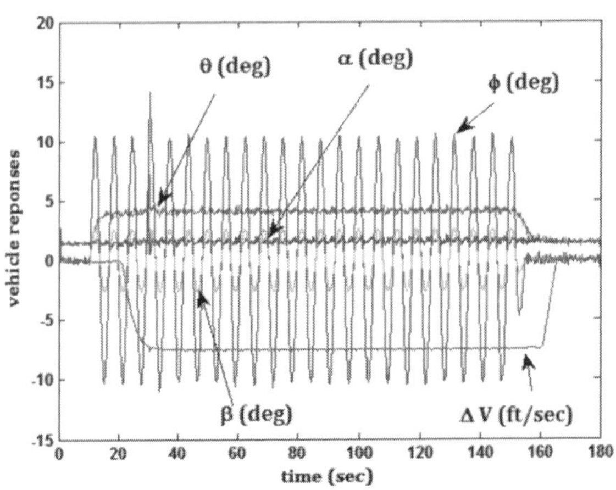

Figure 17. Vehicle responses for Task 5.

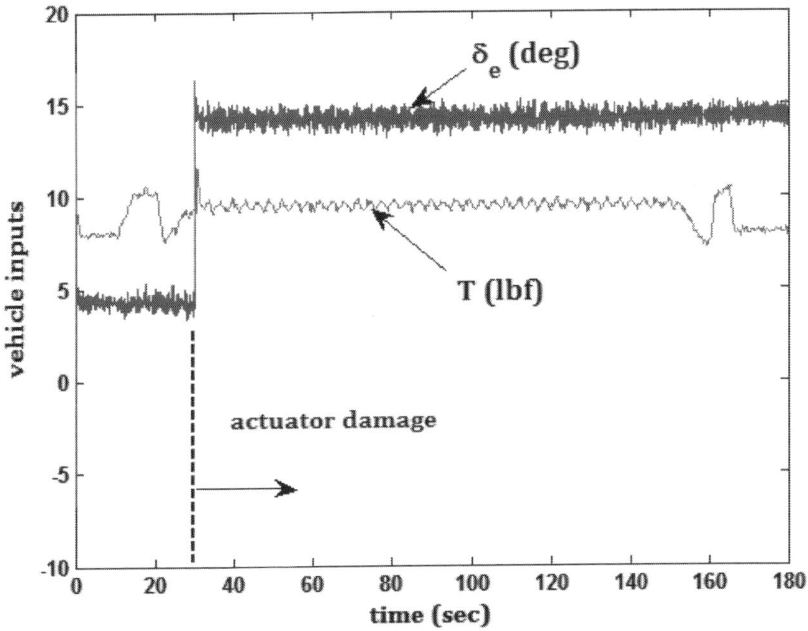

Figure 18. Elevator and thrust inputs for Task 5.

6. DISCUSSION

Quoting from the Introduction: The philosophy behind the approach espoused herein is that the preservation of fundamental pilot input—system response characteristics throughout the flight envelope is important, indeed vital, in preventing LOC events. The computer simulations of Section 5 were intended to provide limited evidence that the SMC design approach has allowed preservation of the fundamental system response characteristics for a number of challenging flight tasks including actuator damage.

With two exceptions, the compensation elements that have been employed in the design were, at most, PI controllers or first-order bandwidth limited rate controllers. The two exceptions were the "hedge dynamics" for the pitch-rate and roll-rate loops. The form of these latter elements, however, is explicitly called out in the design Step (8) of Section 3. The only system "adaptation" that is needed is the scheduling of the observer with flight condition, *i.e.*, altitude and airspeed (Mach No.). This would imply that civil certification of such a design may be considerably less involved than that for a truly adaptive system. As in previous applications of the methodology, control activity is the price to be paid for the robustness that has been demonstrated in the computer simulations. Sensor noise is the primary source of this increased control activity.

7. CONCLUSIONS

Based upon the study summarized herein, the following conclusions can be drawn:

1) The frequency-domain based pseudo-sliding mode design procedure introduced previously in the literature can be successfully applied to the design of a highly nonlinear model of the GTM aircraft.

2) The sliding-mode design can be limited to the inner-most loops of a square control architecture in which nested control loops are closed in sequential fashion.

3) Simple loop-shaping principles are used in each of the loop closures.

4) The majority of the compensators obtained in the design are simple in form, *i.e.*, PI controllers.

5) Computer simulations of the vehicle and control system exhibited stability and performance robustness in different flight conditions and including simulated actuator damage.

6) The overall goal of the study, to demonstrate that fundamental pilot input—system response characteristics could be preserved, with significant variations in flight condition and actuator characteristics, was met.

CONFLICTS OF INTEREST

The author declares no conflict of interest.

REFERENCES

1 Newman, R. Thirty Years of Airline Loss-of-Control Mishaps. In Proceedings of the AIAA Modeling and Simulation Technologies Conference, Minneapolis, MN, USA, 13–16 August 2012.

2 The Boeing Company. Statistical Summary of Commercial Jet Airplane Accidents Worldwide Operations 1959-2008. July 2009. Available online: http://www.airsafe.com/events/ models/statsum 2008.pdf (accessed on 23 October 2013).

3 Wong, C.; Gregory, I.; Cao, C. L1 Adaptive Control with Output Constraints Applied to Flight Envelope Limiting. In Proceedings of the AIAA Modeling and Simulation Technologies Conference, Minneapolis, MN, USA, 13–16 August 2012.

4 Belcastro, C.; Kwatny, H.; Balas, G. Validation of Safety-Critical Systems for Aircraft Loss-of-Control Prevention and Recovery. In Proceedings of the AIAA Modeling and Simulation Technologies Conference, Minneapolis, MN, USA, 13–16 August 2012.

5 Cunningham, K.; Foster, J.V.; Morelli, E.A.; Murch, A.M. Practical Application of a Subscale Transport aircraft for Flight Research in Control Upset and Failure Conditions. In Proceedings of the AIAA Atmospheric Flight Mechanics Conference, Honolulu, HI, USA, 18–21 August 2008.

6 Hedrick, J.K.; Gopalswamy, S. Nonlinear flight control design via sliding methods. *J. Guid. Control Dyn.* **1990**, *13*, 850–858.

7 Shtessel, Y.; Buffington, J.; Banda, S. Multiple timescale flight control using reconfigurable sliding modes. *J. Guid. Control Dyn.* **1999**, *22*, 873–883.

8 Levant, A.; Pridor, A.; Gitizadeh, R.; Yaesh, I.; Ben-Asher, J.Z. Aircraft pitch control via second-order sliding technique. *J. Guid. Control Dyn.* **2000**, *23*, 586–594.

9 Singh, S.N.; Steinberg, M.L.; Page, A.B. Nonlinear adaptive and sliding mode flight path control of F/A-18 model.*IEEE Trans. Aerosp. Electron. Syst.* **2003**, *39*, 1250–1262.

10 Hess, R.A.; Wells, S.R. Sliding mode control applied to reconfigurable flight control design. *J. Guid. Control Dyn.* **2003**,*26*, 452–462.

11 Xu, H.; Mirmirani, M.D.; Ioannou, P.A. Adaptive sliding mode control design for a hypersonic flight vehicle. *J. Guid. Control Dyn.* **2004**, *27*, 829–838.

12 Hess, R.A.; Vetter, T.K.; Wells, S.R. Design and evaluation of a damage-tolerant flight control system. *J. Aerosp. Eng.* **2005**, *219*, 341–359.

13 Hess, R.A. frequency domain-based pseudosliding mode flight control design. *J. Aircr.* **2012**, *49*, 2077–2088.

14 Gregory, I.M.; Xargay, E.; Cao, C.; Hovakimyan, N. L_1 Adaptive Control Law in Support of Large Envelope Modeling Work. In Proceedings of the 1st CEAS Specialists' Conference on Guidance, Navigation & Control, (EuroGNC 2011), Munich, Germany, 13-15 April 2011; pp. 1210–1217.

15 Kwatny, H.G.; Dongmo, J.-E.T.; Chang, B.-C.; Bajpai, G.; Yasar, M.; Belcastro, C. Nonlinear analysis of aircraft loss of control. *J. Guid. Control Dyn.* **2013**, *36*, 149–162.

16 McRuer, D.T. *Aviation Safety and Pilot Control: Understanding and Preventing Unfavorable Pilot-Vehicle Interactions*; National Academy Press: Washington, DC, USA, 1997.

17 Gibson, T.E.; Annaswamy, A.M.; Kenny, S.P. *Adaptive control of Generic Transport Model with Modelled Failures*; ACC/TM-2009-01; Active Adaptive Controls Laboratory, Massachusetts Institute of Technology: Cambridge, MA, USA; January; 2009; Available

online: http://www.mit.edu/~tgibson/AAC200901.pdf (accessed on 23 October 2013).

18 Gorder, P.J.; Hess, R.A. Sequential loop cl*osure in design of robust rotorcraft flight control systems*. *J. Guid. Control Dyn.* **1997**, *20*, 1235–1240.

19 Hess, R.A.; Gama, G. Frequency domain, pseudo-sliding mode control system for a flexible aircraft. *J. Aerosp. Eng.* **2007**, *221*, 707–717.

20 Schmidt, D.K. *Modern Flight Dynamics*; McGraw-Hill: NewYork, NY, USA, 2012; Chapter 10.

21 Heffley, R.K.; Jewell, W.F. *Aircraft Handling Qualities Data*; NASA Contractor Report, NASA CR-2144; NASA: Washington, DC, USA; December; 1972.

22 Hess, R.A.; Cama, G. Flight control system design for inherent damage tolerance. *J. Aircr.* **2008**, *45*, 2024–2035.

23 Utkin, V.I.; Guldner, J.; Shi, J. *Sliding Mode Control in Electro Mechanical Systems*; Taylor and Francis: London, UK, 1999; Chapter 1.

24 Hess, R.A. Metrics for the evaluation of pedal force/feel systems in transport aircraft. *J. Aircr.* **2008**, *45*, 651–662.

25 Gregory, I.M.; Xargay, E.; Cao, C.; Hovakimyan, N. Flight Test of L_1 Adaptive Control Law: Offset Landings and Large Flight Envelope Modeling Work. In Proceedings of the AIAA Guidance, Navigation, and Control Conference, Portland, OR, USA, 8–11 August 2011.

CHAPTER 3

Combining Symbolic Tools with Interval Analysis. An Application to Solve Robust Control Problems

Inès Ferrer-Mallorquí, Josep Vehí

Institut d'Informàtica i Aplicacions, Universitat de Girona, Girona, Spain

ABSTRACT

Complex systems are often subjected to uncertainties that make its model difficult, if not impossible to obtain. A quantitative model may be inadequate to represent the behavior of systems which require an explicit representation of imprecision and uncertainty. Assuming that the uncertainties are structured, these models can be handled with interval models in which the values of the parameters are allowed to vary within numeric intervals. Robust control uses such mathematical models to explicitly have uncertainty into account. Solving robust control problems, like finding the robust stability or designing a robust controller, involves hard symbolic and numeric computation. When interval models are used, it also involves interval computation. The main advantage using interval analysis is that it provides guaranteed solutions, but as drawback its use requires the interaction with multiple kinds of data. We present a methodology and a framework that combines symbolic and numeric computation with interval analysis to solve robust control problems.

KEYWORDS

Interval Analysis, Robust Control, Symbolic Computation

1. INTRODUCTION

The description of real systems using mathematical models has been usually concerned with the concept of uncertainty. The inaccuracies and uncertainties can be done by different causes: in the linearization of a system, for the fact that some parameters of the system can have different values for different operation conditions or because some parameters can vary with time. Moreover uncertainties are classified into two types: parametric or non-parametric. In case the origin and the structure of the uncertainty were known, they are called structured or parametric, otherwise, if they are unknown, the uncertainty is classified as unstructured or non-parametric. From a theoretical point of view, there are a lot of approaches related with parametric robust problems. Most of them use sparametric methodologies [1] -[3] . Some using QFT [4] , other Value Sets [5] .

Some of these methods have been integrated into packages as Matlab [6] or into frameworks as PARADISE [7] . All the cited approaches, manage these problems of robust control using parametric methods. Although these methods have the drawback that the solutions obtained are no guaranteed solutions.

Methods for assessing robust stability and robust design of parameter dependent linear systems, as Routh table or the small gain theorem,... give sufficient (but not always necessary) conditions to solve them. These theorems give a powerful methodology to transform robust control problems in simplified problems consisting of testing the positivity of a set of functions or conditions.

This paper presents a methodology that solves robust control problems combining symbolic computation with interval analysis [8] techniques. The main advantage of interval analysis is that it offers guaranteed solutions in case we want to solve robust control problems. Thus, its use means a really improvement in front parametric techniques.

Although the advantage of obtain guaranteed results is important, the rules and algorithms that must be generated to solve robust control problems using interval analysis, are difficult to be implemented for users. This difficulty discourage control engineers to use these techniques. This difficulty motivates us to offer the users an automatic application that uses interval techniques in a transparent way for the user. This application or framework makes a combination between symbolic tools and interval analysis methods. This paper presented his framework named IRCAD (Interval Robust Control framework for Analysis and Design) dedicated to the resolution of robust control problems applying techniques of interval analysis. This framework is composed by a set of tools of analysis and design that combine symbolic computation with the methodology of interval analysis. Thus, it is not necessary that the user have any special knowledge about interval analysis to use the functions of robust analysis and robust design that offer the framework.

The paper is organized as follows. Section 2 presents the formulation of the most typical robust parametric control problems. The use of interval analysis theory as a way to obtain guaranteed results on the solution of robust parametric control problems is the central point of Section 3. The problem of how to integrate all this diversity of data and all the involved methodologies of this kind of problems is discussed on Section 4. In Section 5 IRCAD is presented as a framework that allows the integration of the different data. Section 6 summarizes the framework and applies it to an illustrative example. Finally most emphasizing points and future intentions are put on the Section 7 of conclusions.

2. ROBUST PARAMETRIC CONTROL PROBLEM

To formulate the problem of parametric robust control we consider a system that is linear respect the variable s. This kind of systems can be expressed as a transfer function or in space state format, in continuous or in discrete domain.

In this paper we will consider a class of plants [9] with structured parametric uncertainties described by the following uncertain transfer function:

$$G(s,q)=\frac{\alpha_0(q)+\alpha_1(q)s+\cdots+\alpha_m(q)s^m}{\beta_0(q)+\beta_1(q)s+\cdots+\beta_n(q)s^n} \tag{1}$$

depending on a structured perturbation characterized by the parameter vector

$$\boldsymbol{q}=[q_1 \quad q_2 \quad \cdots \quad q_l]^{\mathrm{T}} \tag{2}$$

where each parameter enters into the system description with polynomial dependency. We also consider a certain configuration of the feedback system with a fixed controller $C(s,\boldsymbol{k})$, where $\boldsymbol{k}$ is the design parameter vector. The motivation of this kind of system description is that the system parameters, q, can represent physical quantities that are known only to within a certain accuracy, or vary depending on operating conditions while the controller parameters, $\boldsymbol{k}$, represent degrees of freedom available to the control system designer. Due to the physical interpretation of the uncertain parameters, each one can be considered independent from the other and their values lie between upper and lower bounds. Then, the uncertain domain can be defined as any hyperrectangle:

$$\boldsymbol{Q}=\left\{\boldsymbol{q}=[q_1 \quad q_2 \quad \cdots \quad q_l]^{\mathrm{T}} \,\middle|\, q_i \in \left[q_i^-,q_i^+\right], i=1,\cdots,l\right\} \tag{3}$$

With this kind of feedback system description, several robust control problems can be formulated. Problems concerning robust control analysis as testing the stability of the system or computing the stability margin. Also, problems related with control design as to find a set of controllers that fulfill one or more than one selected specifications.

2.1. Robust Control Problems: Analysis

In many control systems the plant parameters may vary over a wide range about a nominal value p^0. Robust parametric stability refers to the ability of a control system to maintain stability despite such large variations. Given an uncertain system and a controller, analysis of robust control problems consists in to check(calculate) some performance specifications:

Robust stability

The first problem formulated is the verification of the stability on uncertain systems. That is, how to find the necessary and sufficient conditions that allows to guarantee that a family of uncertain polynomial is stable or no. In [10] , Ackermann presents different methods to check robust stability analysis.

-Boundary crossing theorem: It is based on Hurwitz determinants. Given a set of polynomials $P(s,Q)$, this set is robustly stable, if and only if:

A) There exists a stable polynomial $p(s,q) \in P(s,Q)$ 2) Any root cross imaginary edge: $jw \notin \text{Roots}\left[P(s,Q)\right]$ for all $\omega \geq 0$

-Alternative boundary crossing theorem based on Hurwitz Determinants [11] : Giving a set of polynomials $P(s,Q)$ it is robustly stable, if and only if:

1) There exists a stable polynomial $p(s,q) \in P(s,Q)$

2) $\det H_n(q) \neq 0$ for all $q \in Q$

This work uses the just presented boundary crossing theorem in a modified way. It gives necessary and sufficient conditions of absolute robust stability for a family of characteristic polynomials.

Stability margin

If the controller is given, the parametric stability margin is defined [12] as the maximal range of variation of the parameter $\boldsymbol{p}$, measured in a suitable norm, for which closed-loop stability is preserved. It can be expressed as:

$$\rho_x := \sup\left\{\alpha : \delta\left(s, \boldsymbol{x}, \boldsymbol{p}\right) \text{ stable}, \left\|\boldsymbol{p} - \boldsymbol{p}^0\right\| < \alpha\right\} \tag{4}$$

This margin is used as a quantitative measure of the robustness of the closed loop system with respect to parametric uncertainty evaluated at the nominal point p^0.

Parametric bode plots

Based on the experience with frequency domain analysis methods, one might conjecture that it is not necessary to check all $\boldsymbol{\omega}$ values. Thus, it is common to construct Bode plots using only a grid of $\boldsymbol{\omega}$ values. In the case of frequency domain methods for robustness analysis, the bottleneck is "singular frequencies".

In order to provide a systematic method of identifying singular frequencies, an algebraic definition is given on [10] . Before the definition some preliminary notation is introduced. When an uncertain polynomial $p(s,\boldsymbol{q})$ is evaluated at a frequency $\boldsymbol{\omega}$ it equals a complex number $p(j\omega,\boldsymbol{q}) = h(-\omega^2,\boldsymbol{q}) + j\omega g(\omega^2,q)$. Separating this complex number on real and imaginary parts:

$$h(-\omega^2,\boldsymbol{q}) = a_0(\boldsymbol{q}) - a_2(\boldsymbol{q})\omega^2 + a_4(\boldsymbol{q})\omega^4 - \text{L} \tag{5}$$

$$\omega g\left(-\omega^2, q\right) = a_1(q)\omega - a_3(q)\omega^3 + a_5(q)\omega^5 - \text{L} \tag{6}$$

Associated with these functions is a Jacobi an matrix.

$$J(\omega,\boldsymbol{q}) = \begin{pmatrix} \dfrac{\delta h\left(-\omega^2,\boldsymbol{q}\right)}{\delta q_1} & \dfrac{\delta h\left(-\omega^2,\boldsymbol{q}\right)}{\delta q_2} & \text{L} & \dfrac{\delta h\left(-\omega^2,\boldsymbol{q}\right)}{\delta q_l} \\ \dfrac{\delta \omega g\left(-\omega^2,\boldsymbol{q}\right)}{\delta q_1} & \dfrac{\delta \omega g\left(-\omega^2,\boldsymbol{q}\right)}{\delta q_2} & \text{L} & \dfrac{\delta \omega g\left(-\omega^2,\boldsymbol{q}\right)}{\delta q_l} \end{pmatrix}$$

In the Jacobian $J(\omega,\boldsymbol{q})$, the real part $h\left(-\omega^2,\boldsymbol{q}\right)$, and the imaginary part $\omega g\left(\omega^2,q\right)$ are used to give algebraic definition of singular frequencies.

Definition. The nonnegative frequency ωs is a singular frequency of the uncertain polynomial $p(s,\boldsymbol{q})$ if there exists a $\boldsymbol{q}^0 \in \mathrm{R}^l$ such that the three following conditions are simultaneously satisfied.

$$h\left(-\omega_s^2,\boldsymbol{q}^0\right) = 0 \tag{7}$$

$$\omega_s g\left(-\omega_s^2,\boldsymbol{q}^0\right) = 0 \tag{8}$$

$$rank\left[J\left(\omega_s,\boldsymbol{q}^0\right)\right] < 2 \tag{9}$$

The last Equation (9) can be substituted by: $\det \boldsymbol{J}(\omega\boldsymbol{q}) = 0$

Frequency problems

-Value sets. Some gridding approaches have as a drawback the length of time it takes to compute the set of frequency plots. An alternative and faster technique, suggested in [5] , is to compute the frequency plot $p(j\omega,q),\omega \geq 0,$ for each q on a grid of Q. It is advisable to compute the value set for each ω on a grid of frequencies from 0 to +∞.

$$P(j\omega,Q)=\{p(j\omega,q)\in \mathbb{C}\,|\,q\in Q\} \tag{10}$$

In the case of a family of polynomials, the stability test is based on repeating the construction of the value set for a grid of frequencies to give the set of all possible frequency plots. The collection of value sets would then indicate stability or instability depending on how the zero-exclusion theorem is formulated.

Theorem 1. (zero-exclusion theorem). Given a polynomial family $P(s,Q)=\{p(s,q)|q\in Q\}$. This set is robustly stable if and only if

- A stable polynomial $p(s,q)\in P(s,Q)$ exists and
- $0\notin P(j\omega,Q)$ for all $\omega \geq 0$.

Discrete-time problems

Some problems concerned with discrete-time are:

-Model conversion. Conversion is needed in both senses, from a continuous time interval state-space model to a discrete-time interval model and also from a discrete-time uncertain system to a continuous-time uncertain model.

-Schur stability test. The problem of checking the stability of a discrete-time system is reduced to the determination of whether or not the roots of the characteristic polynomial of the system lie strictly within the unit disc, that is whether or not the characteristic polynomial is a Schur polynomial. If

$$P(z) = p_n z^n + p_{n-1} z^{n-1} + \text{L} + p_1 z + p \quad (11)$$

where the z_i are the n roots of $P(z)$. Then if $P(z)$ is Schur, all these roots are located inside the unit circle $|z| < 1,$ so that when z varies along the unit circle $z = e^{j\theta}$, the argument of $P(e^{j\theta})$ increases monotonically. For a Schur polynomial of degree n, $P(e^{j\theta})$ has a net increase of argument of $2\pi n$, and thus the plot of $P(e^{j\theta})$ encircles the origin n times. This can be used as frequency domain test for Schur stability [13] [14] .

-Non-linear discrete-time control of uncertain systems. This problem can be formulated as:

Find one c

$$c \in S_c = \{c \in C | \forall p \in P, f(c,p) > 0\} \quad (12)$$

where f is a vector function that can be evaluated using algorithms based on interval analysis. This problem is both quite complex because it involves a quantifier and, at the same time, very simple because the only objective is to find one feasible vector. This makes considering a larger number of tuning parameters corresponding to the guaranteed-tuning problem [15]. This is evidence that it is possible to combine non-linearity, and structured uncertainty with guaranteed results.

2.2. Robust Control Problems Design

The requirement of robust design is that the design specifications must be satisfied over the entire parameter set. Given an uncertain system, the problem is to obtain a robust controller that fulfills some performance specifications.

As robust performance specifications we can enumerate stability, pole locations, steady-state, H_∞, etc. It is known that the solution of these problems can be reduced to the problem of checking the positivity of a set of rational functions over a given domain:

$$f_i(\alpha, \boldsymbol{q}, \boldsymbol{k}) > 0 \quad \forall \alpha \in A \quad \forall \boldsymbol{q} \in \boldsymbol{Q} \quad \forall \boldsymbol{k} \in \boldsymbol{K} \tag{13}$$

where α is the generalized frequency and $\boldsymbol{k} \in \boldsymbol{K}$ may be a single point (nominal controller $\boldsymbol{k}^0$) or a certain domain in the parameter's space of the controller depending on the problem considered.

Ÿ Robustness analysis: to check if the controlled system achieves robust performances.

$$f_i(\alpha, \boldsymbol{q}) > 0 \quad \forall \alpha \in A \quad \forall \boldsymbol{q} \in \boldsymbol{Q} \tag{14}$$

Ÿ Robust control design: to find the robust K set such that the controlled system achieves robust performances.

$$f_i(\alpha, \boldsymbol{q}, \boldsymbol{k}) > 0 \quad \forall \boldsymbol{k} \in \boldsymbol{K} \quad \forall \boldsymbol{q} \in \boldsymbol{Q} \tag{15}$$

Most of the more used robust control design techniques are based on the worst possible scenario which may never occur in a particular control system. Some of these techniques are the following:

Ÿ H_∞ control [16] . Robust performance specifications may be given as pole locations as well as H_∞ performances expressed in the frequency domain.

Ÿ l_1 control [17] . In the standard l_1 problem the design of an internally stabilizing controller such that the l_1 norm of the regulated output z due to the worst-case magnitude bounded disturbance ω is addressed. The Banach space of right-sided absolutely summable real sequences with the norm given by

$$\|x\|_1 := \sum_{k=0}^{\infty} |x(k)| \tag{16}$$

Ÿ μ synthesis. It minimizes the superior bound of the robust performance. The criterion is

$$J = \min_{t_p} \left\| \left[\min_{c(s)} \left\{ \sup \mu_{A(s)} \left[M(s) \right] \right\} \right] - 1 \right| \tag{17}$$

where $\mu[0]$ is the structured singular value, $A(s)$ is the disturbance matrix which indicates that $\mu[0]$ depends on the allowed structure for $A(s)$, $M(s)$ is a matrix obtained by rearranging the uncertain system into the M A-structure and $C(s)$ indicates that the minimisation is function of the controller parameters [18] [19] .

Ÿ QFT (Quantitative feedback theory). While value sets are mainly concerned with analysis, when the aim is finding a compensator to satisfy design specifications, QFT is the most important approach. It can be considered as a natural extension of classical frequency-domain design approaches. One of the main objectives is to design a simple low-order controller where the bandwidth of the feedback controller being as small as possible. At a fixed frequency, the plant's frequency response set is called a template. In the bound generation step of QFT design procedure, the plant template is used to translate the given robustness specifications in domains in the Nichols chart where the controller gain-phase values are allowed to lie.

We are focused on the solution of robust control design problems. That is, we are testing if a controller K achieves the design specifications.

Some applications can be found in the design field. Parameter spaces of the controller are explored, finding the regions which fulfill a control specification in order to obtain a controller which satisfies it.

3. INTERVAL FORMULATION OF ROBUST PARAMETRIC CONTROL

The problem of robust parametric control, presented on the last section, is now formulated using interval analysis. Using the theory of interval analysis is possible to obtain guaranteed solutions to robust control problems. To formulate these problems on interval format, we formulate the problem of the parameter space approach considering the uncertain system of Figure 1, where k is the parameter vector of the controller,

$$\boldsymbol{k} = \begin{bmatrix} k_1 & k_2 & \mathrm{L} & k_l \end{bmatrix}^{\mathrm{T}} \tag{18}$$

q is the parameter vector of the process, and the uncertainty domain can be defined as a box:

$$K = \left\{ \boldsymbol{k} = [k_1 \quad k_2 \quad \ldots \quad k_l]^{\mathrm{T}} \,\middle|\, k_i \in \left[\underline{k_i}, \overline{k_i}\right] \right\} \tag{19}$$

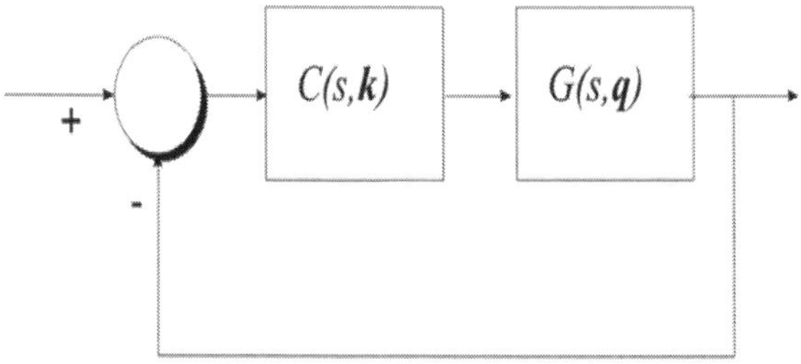

Figure 1. Uncertain system.

Design specifications are formulated in terms of closed-loop system stability and performances in the frequency domain. These specifications, such as bandwidth, resonance peak, control effort, etc., can be described as a set of N inequalities of the type:

$$f_i(\omega, \boldsymbol{q}, \boldsymbol{k}) > 0 \quad \omega \in \Omega \quad \boldsymbol{q} \in \boldsymbol{Q} \quad \boldsymbol{k} \in \boldsymbol{K} \quad i = 1, \ldots, N \tag{20}$$

where Ω is a subset of $\mathbb{R}^+$ (usually an interval), Q an interval vector representing interval parameters and K; a nondegenerate set.

Then robust design can be formulated as follows: given a controller structure, C(s, K) the aim is to find the values of k which conform with the robust control specifications. Taking this formulation into account some control problems can be proposed [20] :

1) Performance checking. Given the uncertain system $G(s, \boldsymbol{Q})$ and the uncertain domain $\boldsymbol{Q}$, the problem is to test if the designed controller $C(s, \boldsymbol{k}^0)$ achieve the robustness specifications obtained from Equation (20):

$$F_i\left(\omega,\boldsymbol{Q},\boldsymbol{k}^0\right)>0 \tag{21}$$

2) Performance margin computation. Taking set $\prod$ as a function of its radius ρ:

$$\prod(\rho)=\left\{\boldsymbol{q}:\left\|\boldsymbol{q}-\boldsymbol{q}^0\right\|_\infty^\omega\leq\rho\right\} \tag{22}$$

find the maximal set $\prod(\rho^*)$ so that the designed controller $C\left(s,\boldsymbol{k}^0\right)$ achieves robust performances for all ρ belonging to $\prod(\rho^*)$.

3) Robust controller design. Taking a particular structure for the controller and specifying the uncertain domain where the parameters of the system $\boldsymbol{Q}$ can vary, find a fixed structure controller $C\left(s,\boldsymbol{k}^0\right)$ so the controlled closed loop system achieves the robust performance specifications:

$$F_i\left(\omega,\boldsymbol{Q},\boldsymbol{k}^0\right)>0 \tag{23}$$

4) Obtaining the set K for the robust controller problem. Given an uncertain plant $G(s,\boldsymbol{q})$ the variation domain of the system parameter $\boldsymbol{Q}$ and a controller structure, find the robust set $\boldsymbol{K}$ which allows $C(s,\boldsymbol{k})$ to achieve the robustness specifications:

$$F_i\left(\omega,\boldsymbol{Q},\boldsymbol{k}\right)>0 \tag{24}$$

5) Estimating the stability region problem, for a given k^0. This problem consists on building a set of all ρ 's achieves closed loop stability when given k^0.

In our approach two robust control problems, Design and Analysis, have been converted into a set of functions using symbolic computation. These functions have been intervalised resulting on a set of interval functions. At this point, to solve the problems it has only been necessary to test the positivity of these interval functions for all the variation of its parameters, Equation (14) and Equation (15).

4. INTEGRATION OF NUMERIC, INTERVAL AND SYMBOLIC COMPUTATION

As it has seen in Section 2, problems related with robust parametric control systems use symbolic data due these systems involves uncertain variables. Moreover in Section 3 is denoted that the interval formulation of the problems involves numeric and interval data. Thus, solving robust control problems using interval analysis tools, offer guaranteed results, but as a drawback, its use requires the interaction with multiple kind of data. These difficulties can discourage control engineers to use interval techniques although its benefits.

To make all these tools and techniques more available is interesting an environment that integrates the treatment of all these kind of data, allowing deal with robust control problems, solving them using a common methodology.

In this sense our work has found how to offer to the control engineer a set of friendly user interface tools that integrates the diversity of data that these kinds of problems involves and also allows to use interval tools in a transparent way for the user.

In Figure 2, a schema of the different tools used for these kinds of problems to manage numeric symbolic and interval computation is shown.

As it can be seen in this schema, Matlab is the package used to centralize our environment. It is the selected package to develop the most routines of our environment due three main points:

- The high computational power with problems that involves matrix data.

- Its facility to deal with high level languages.

- Its easy connection with other commercial packages as Maple.

Our framework solves robust control problems, thus, problems that have implicitly high level degree of computation. This aspect would be a bottleneck if they would be solved writing scripts on a package like Matlab. Due this high cost of computation the routines on this framework have been implemented with the high level language C++. This language improves the building of procedures which require numeric-interval computation and allows to execute code in a faster way than using Matlab. This achievement was very decisive. In addition it allows to use tools from interval analysis libraries [21] , due its include instructions.

Once presented the two main packages used by the framework, it rests to say how Matlab procedures link with C++ critical routines. The deal of procedures and data among these two packages is done using code CMEX and DLL functions. The high level C++ functions are compiled using Borland, obtaining a set of dll function. There are executable functions that can be called from Matlab commands or from Matlab scripts.

Another aspect to consider is how to enter symbolic data, such as transfer functions or problem specifications. In our case it has been choose the Maple package. The extraction of these symbolic inputs from Matlab scripts and the pass of them to the Maple Package are implemented using the Symbolic Math Toolbox.

Therefore, Matlab is used to make the integration of numeric, interval and symbolic data, taking symbolic inputs, transforming them into a suitable format for the C++ routines, passing these transformed inputs to the C++ routines and finally collecting the data returned.

The last step, shown in the lower part of the diagram (user interface block) corresponds to the need to show results in a suitable way for control applications. The framework has implemented this part using the GUI-tools of Matlab. These tools allow to achieve a friendly interface that allows to present the results in an adequate format to be analyzed easily by the user.

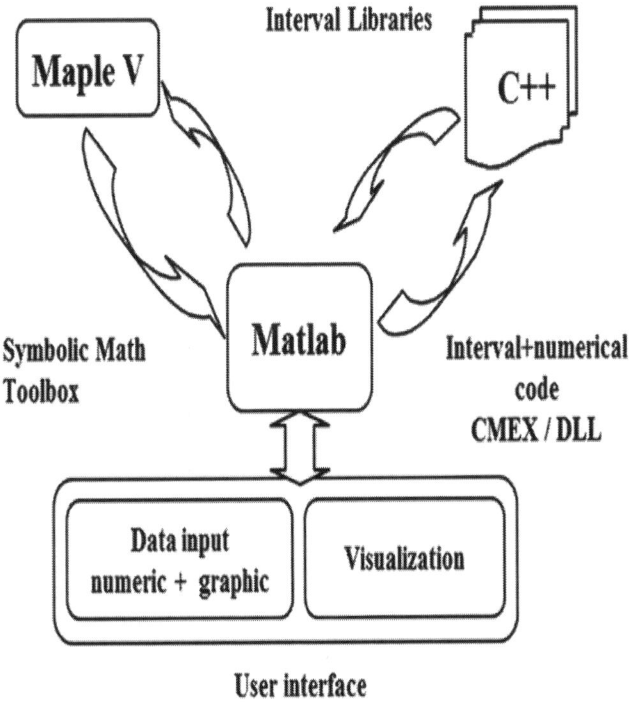

Figure 2. Cooperation numeric-interval-symbolic.

5. IRCAD: A FRAMEWORK FOR ROBUST CONTROL PROBLEMS

Once introduced the need to integrate the diversity of data and methods in section 4, now in this section we present the developed environment in order to achieve this integration. To obtain the kindly manipulation of numeric, interval and symbolic computation, our work has built a set of tools under the format of a framework called IRCAD (Interval Robust Control framework for Analysis and Design). To introduce the framework IRCAD (main screen is showed on Figure 3), we present the functions that it integrates. These functions or tools could be laid as three sets of problems.

The first set includes robust analysis problems like to test the stability of a system and to find the stability margin. The second set of problems are related to robust design problems and the main tool is an application that allows to select the design specifications that we want our control system achieve. The framework includes a third set corresponding to post-design tools, which allows that a selected controller can be applied to the system without leave the framework IRCAD.

The starting point to use the tools of the framework is to input the transfer function of the plant. Due the systems we consider are uncertain systems, the plant is defined as a family of transfer functions. To input them into the framework it is necessary to define a transfer function with undefined parameters, and also define their uncertain values range (interval values). So, as example, we can define a transfer function for a family of plants like $G(s,q)$ with the uncertain parameter $q=[q_1,q_2]$.

5.1. Tools for Robust Analysis

The framework IRCAD has a set of tools addressed to solve robust analysis problems. Giving an uncertain system, where the transfer function is defined as a family of plants q_i, from the menu Analysis, it is possible to make the following functions:

-Stability test. To calculate the absolute stability region, a branch-and-bound algorithm based on interval analysis is used. The result of this tool shows the regions stable and unstable. In case the depth of the algorithms was limited (in order to accelerate the end of the computation) can appear undefined regions. In the case there aren't problems of time computation, due the computer has high computation resources, it is possible to don't limit the depth of search and then the lonely regions that appear are estable and unestable regions.

-Stability margin computation. The parametric stability margin is defined as the length of the smallest perturbation Δq which destabilizes the closed loop. This margin is used as a quantitative measure of the robustness of the closed loop system with respect to parametric uncertainty evaluated at the nominal point q_0.

-Parametric Bode plots. It is common to construct Bode plots using only a grid of ω values. There are, however, cases when 'singular frequencies' occur. The framework considers these cases which require special attention in all frequency domain methods for robustness analysis.

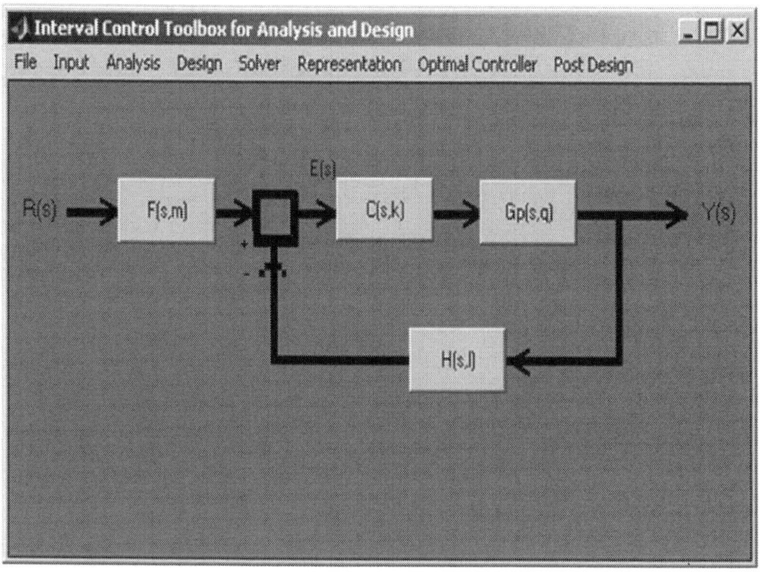

Figure 3. IRCAD framework.

5.2. Tools for Robust Design

The IRCAD framework has a powerful set of tools aimed at the design of controllers in the case of parametric control systems. Given an uncertain system, where the transfer function is defined as a family of plants q_i, from the menu design it is possible to select one or more than one specification. The specifications involve temporal and frequency domain and include: stability degree, absolute stability, resonance peak, velocity error, settling time, control effort and overshoot.

Once the specifications have been selected, each chosen specification is translated to an interval function, converting them into matrix structures and passing them to a solver (a C++ routine). The results of the solver routine, that is to say the results of the design problem, are returned on a matrix format as a set of controllers that fulfill the selected specification or the set of specifications. The resultant controllers are a set of interval values. Using the tools of Graphic User Interface from Matlab (GUI tools), the resultant matrix, which contains all the information about the valid controllers is passed to the user.

The framework allows to choose in which format we want to show the set of valid controllers. It is possible to visualize this set in a graphic way or in a numeric way. The decision often depends on the number of parameters selected for the controller. If the user wants to design a controller of two or less parameters, he can choose the graphical format, then the result will be a graph with the parameter space ranged by the parameters of the controller (k_1, k_2). In this graph the set of valid controllers are colored as a red region (Figure 5), the unfeasible controllers on a yellow region and the undefined regions on blue regions. If the user choose to obtain the set of valid controllers in a numeric format, then the framework shows on a table, the set of intervals for k_1, k_2, L, k_i that allows the system fulfill all the specifications.

5.3. Post-Design Tools

To help control engineer, IRCAD framework offers a very complete and power fulset of post-design functions. These functions allow the user, on the one hand to complete the design selecting one interval controller from the feasible controller set and in the other hand allow to verify that the selected controller achieve the specifications, without exit from application.

In order to offer a more reliable framework, the package gives the possibility to choose a criterion to select the optimal controller.

-Optimal Controller. The results of the robust control problem design are returned as a set of controllers that fulfill the specification or the set of specifications selected. To choose the optimal controller among this set is a task that can be not easy for the user, although the user was control engineer. To make the selection of the optimal controller from the given set of feasible controllers computed by the framework IRCAD, that offers the possibility to do it using three different criteria:

Ÿ Minimum norm: Given a graphical region of feasible controllers, in the parameter space delimited by the parameters of the controller K_1 and K_2, this criteria select the interval controller (K_1, K_2) that, fulfilling the specifications, has the minimum norm. This measure is calculated from the origin of the parameter space to the square defined by the interval (K_1, K_2).

Ÿ Maximum robustness: In this case the framework computes over each square (K_1, K_2) the value of the stability margin and selects as controller the square (K_1, K_2) that have the maximum value.

Ÿ Neighboring: Over each feasible square (K_1, K_2), the interval controller, the framework computes the number of neighbors. The square (K_1, K_2) with the maximum number of neighbors is selected.

When the user selects one of these criteria, then IRCAD computes one recommended controller following the method chosen. The framework shows the obtained controller by two ways at time: graphically and numerically. Graphically, emphasizing this controller with a dark color on the graphic, that represents the set of feasible controllers (Figure 6), and Numerically, where is opened a window with the recommended values for the controller.

The recommended value for the controller can be introduced on the controller box as an entry, or it can be introduced automatically as following is explained.

-Input optimal controller. The framework allows that the parameters of the commended controller obtained using one of the techniques exposed (minimum norm, maximum robustness or nearest neighbor) can directly be input. This automatic entry of the parameters of the controller allows to execute post-design tasks without exit from the framework.

-Post design tools. It is not necessary to go to any simulation environment to verify the selection, because the selected controller can be simulated using the same framework IRCAD.

All these features give to the user a friendly environment to work in the case of robust control problems, with the security that he has obtained guarantee solutions on the results.

The mainly tools of this block are concerning to the facilities to have tools to check that a designed controller fulfill a set of specifications.

6. ILLUSTRATIVE EXAMPLE

To illustrate how the solver created manages the data, an example suggested by [22] and later studied by [23] is used. The aim of the example is tuning a PI controller for an interval plant.

Example 1. Robust control design.

Given the transfer function of the plant:

$$G(s,\boldsymbol{q}) = \frac{q_1}{1 - \dfrac{s}{q_2}} \tag{25}$$

where q is the vector of uncertain parameters , $q_1 = q_2 = [0.8, 1.25]$ $\boldsymbol{q} = [q_1 \quad q_2]$; and the PI controller is described by:

$$C(s,k)=\frac{k_1\left(1+\frac{s}{k_2}\right)}{s} \tag{26}$$

where $\boldsymbol{k}$ is the vector of the design parameters, $\boldsymbol{k}=[k_1,k_2]$.

The design aim is to find the set $\boldsymbol{K}$ of controller parameters that fulfill the following performance specifications:

1. Absolute stability.

2. Velocity error lower than 2%.

3. Resonance peak lower than 3 dB.

4. Control effort lower than 20.

Given the Equation (26), the toolbox IRCAD manage each one of these specifications transforming them via the Extended Symbolic Toolbox into a set of polynomial interval functions [Equations (27)-(30)]. Matlab send these functions to the solver, transforming it previously on a matrix structure. This transformation is automatically done by IRCAD toolbox, thus it is transparent to the user. The role of the solver is to test the positivity of these functions over the parameter space, which is determined by the controller parameters. In this example the parameter space is determined by k_1 and k_2. IRCAD allows to present the results of the solver in two formats :

-Graphically.

-Set of intervals.

For this example the parameter space to be checked is limited giving the initial interval values to the parameters of the controller as $k_1=[-200,0]$ and $k_1=[0,10]$. In order to work with positive intervals the change $\overline{k_1}=-k_1$ has been made.

As is exposed at Section 2, in the subsection dedicated to robust control problems of design, to test if one specification or a set of specifications are fulfilled it is enough to test if a function or a set of functions are positive. In our example it is enough to test the positivity of the following functions:

1) For the specification of absolute stability the function obtained is:

$$f_1 = -k_2 + q_1 \overline{k_1} \tag{27}$$

2) For the specification of velocity error less than 2%:

$$f_2 = q_1 \overline{k_1} - 50 \tag{28}$$

3) For the specification of resonance peak lower than 3 dB:

$$f_3 = 2k_2^2\omega_1^4 - 4k_2^2\omega_1^2\overline{k_1}q_1q_2 + \overline{k_1}^2 q_1^2 q_2^2 k_2^2 + 2k_2^2\omega_1^2 q_2^2 - 4k_2\omega_1^2 q_2^2 \overline{k_1} q_1 + \overline{k_1}^2 q_1^2 q_2^2 \omega_1^2 \tag{29}$$

4) And finally for control effort lower than 20:

$$\begin{aligned} f_4 = {} & 400k_2^2\omega_1^4 - 800k_2^2\omega_1^2\overline{k_1}q_1q_2 + 400\overline{k_1}^2 q_1^2 q_2^2 k_2^2 + 400k_2^2\omega_1^2 q_2^2 + 400\overline{k_1}^2 q_1^2 q_2^2 \omega_1^2 \\ & - \overline{k_1}^2 k_2^2 q_2^2 - \overline{k_1}^2 \omega_1^4 - \overline{k_1}^2 k_2^2 \omega_1^2 - \overline{k_1}^2 \omega_1^2 q_2^2 - 800k_2\omega_1^2 q_2^2 \overline{k_1} q_1 \end{aligned} \tag{30}$$

The resultant feasible regions for each specification are represented on Figure 4.

In this example, specification f_3 and specially specification f_4 have a hard computation cost due its high number of summand components and also for the high number of plant variables q_i and controller variables k_i that they involve.

Although this, when IRCAD compute the four specifications at the same time the cost of computation is significantly reduced because some regions are discarded on the specification of lost cost f_1 and f_2.

Using IRCAD, the problem of design a controller that fulfills the four specifications gives the graphical classification of Figure 5.

The graphical result of Figure 5 corresponds to the following set of feasible intervals:

$$k_1 = [-70, -65] \quad k_2 = [2.75, 3]$$

$$k_1 = [-75, -65] \quad k_2 = [3, 3.25]$$

$$k_1 = [-80, -65] \quad k_2 = [3.25, 3.5]$$

$$k_1 = [-85, -65] \quad k_2 = [3.5, 3.75]$$

$$k_1 = [-90, -65] \quad k_2 = [3.75, 4]$$

$$k_1 = [-95, -65] \quad k_2 = [4, 4.25]$$

$$k_1 = [-100, -65] \quad k_2 = [4.25, 4.5]$$

$$k_1 = [-100, -75] \quad k_2 = [4.5, 4.75]$$

$$k_1 = [-105, -80] \quad k_2 = [4.75, 5]$$

$$k_1 = [-110, -85] \quad k_2 = [5, 5.25]$$

$$k_1 = [-110, -95] \quad k_2 = [5.25, 5.5]$$

$$k_1 = [-110, -100] \quad k_2 = [5.5, 5.75]$$

$$k_1 = [-110, -105] \quad k_2 = [5.75, 6]$$

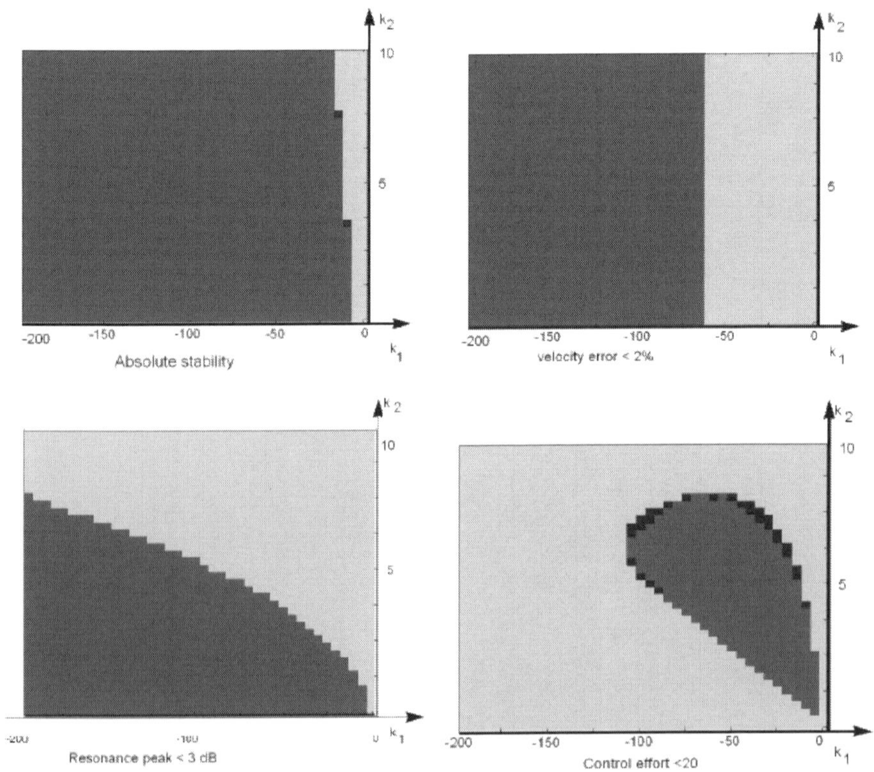

Figure 4. Feasible regions for each specification.

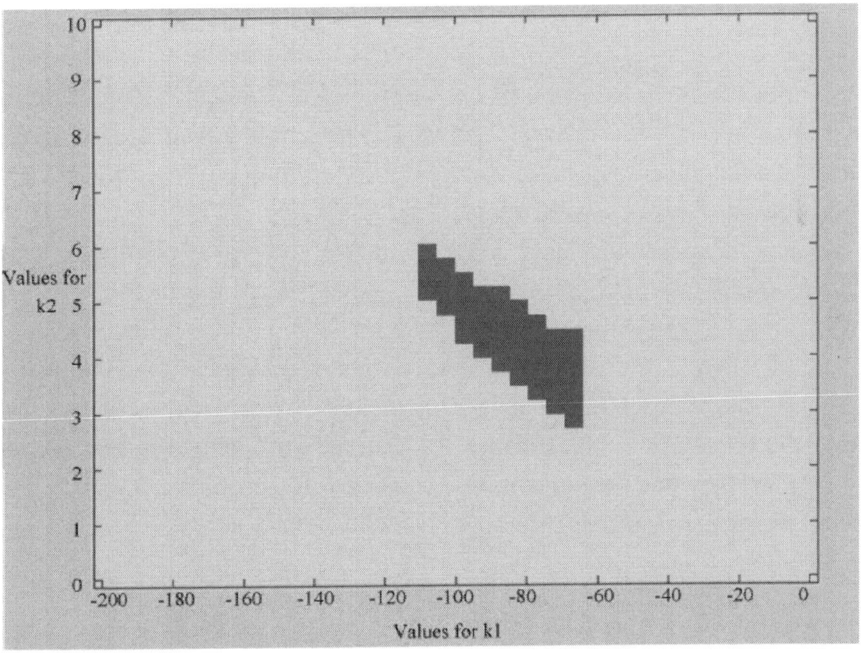

Figure 5. Feasible regions.

Finally, the solution of this robust control design example consists on the choice from this feasible region, following some criterion, of an interval pair k_1 and k_2 for the PI controller.

At this point IRCAD offers a set of post-design tools that allows to choose the optimal controller from the set of the proposed controllers on an automatic way. The framework chooses the controller following the criteria (see Section 5.3) that the user has selected.

In this example we suppose the user select the criterion of minimum norm. Then the controller suggested by the framework IRCAD is $[-67.5, 2.875]$ as is shown on Figure 6.

This action gives the optimal controller that fulfills the set of specifications. Using another post-design tool of the framework, the user can put this optimal controller on the closed-loop system. Finally, with this controller and this plant the framework offers some post design tools to simulate the features of the system.

7. CONCLUSIONS AND FUTURE WORK

An application to solve robust control problems was designed combining symbolic tools with interval analysis. The main advantage of using interval analysis is that guaranteed results are obtained. The drawback is the necessity to work with a variety of data, implicit on these kinds of problems. It is necessary to manage symbolic, interval and numeric data. The manipulation of these kinds of data needs to use the fitting software in each case. This diversity can discourage control engineers because most of them don't use these data in a mixed way. The framework created, allows users to manage on an easy way with problems that involves these kinds of data in a transparent way. The result is a most reliable framework for control engineers to solve problems that involves symbolic-interval-numeric data.

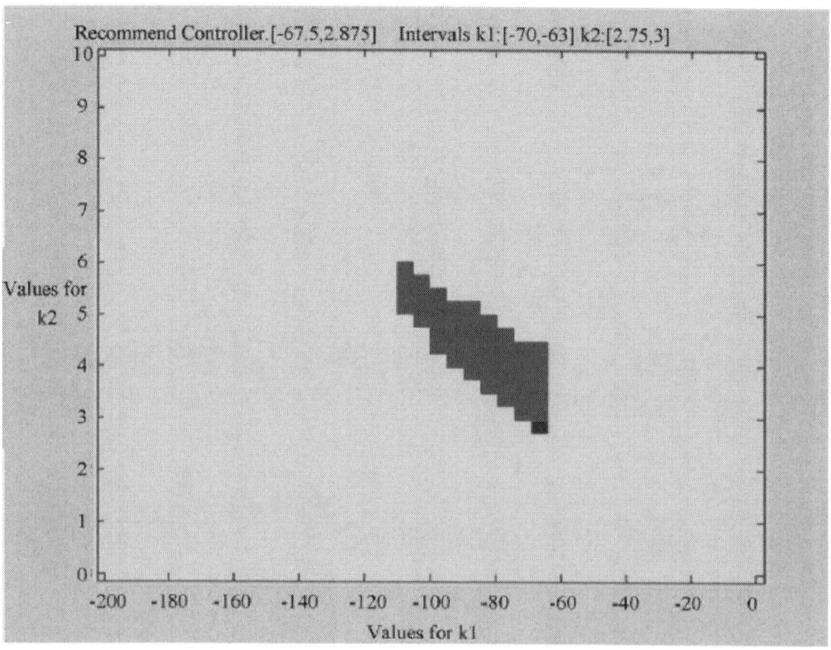

Figure 6. Minimum norm criteria.

REFERENCES

1 The MathWorks Inc. (1999) The Polynomial Toolbox for Matlab, v. 2.0. Technical Report. www.polyx.com

2 Sakabe, K., Yanami, H., Anai, H. and Hara, S. (2004) A Matlab Toolbox for Robustcontrol Synthesis by Symbolic Computation. SICE Annual Conference in Sapporo, Hokkaido Institute of Tecnology, Japan, August 2004, 1968-1973.

3 Sienel, W., Bunte, T. and Ackermann, J. (1996) Paradise: Parametric Robust Analysisand Design Interactive Software Environment: A Matlab-Based Robust Control Toolbox. Proceedings of the 1996 IEEE International Symposium on Computer-Aided Control System Design, Dearborn, 15-18 September 1996, 380-385.

4 Nataraj, P.S.V. and Sardar, G. (2000) Computation of QFT Bounds for Robust Sensitivity and Gain-Phase Margins Specifications. Journal of Dynamic Systems Measurement and Control, 122, 528-834.

5 Ackermann, J. (1993) Robust Control: Systems with Uncertain Physical Parameters. Communications and Control Engineering Series. Springer-Verlag, Berlin.

6 Hyodo, N., Hong, M., Yanami, H. and Anai, H., et al. (2006) Development of a Matlab Toolbox for Parametric Robust Control: New Algorithms and Functions. SICE-ICASE International Joint Conference 2006, Busan, 18-21 October 2006, 2856-2861,

7 Ackermann, J. (2000) PARADISE—Parametric Robustness Analysis and Design Interactive Software Environment. Technical Report, Institute of Robotics and Mechatronics. www.op.dlr.de/FF-DRRR/paradise.

8 Moore, R.E. (1979) Methods and Applications of Interval Analysis. In: SIAM Studies in Applied Mathematics, SIAM, Philadelphia.

9 Vehí, J., Herrero, P., Sainz, M.A. and Jaulin, L. (2004) Quantified Set Inversion with Applications to Control. IEEE International

Symposium on ComputerAided Control Systems Design, Taipei, 2-4 September 2004, 179-183.

10 Ackermann, J. (2002) Robust Control. The Parameter Space Approach. Springer, Berlin.

11 Frazer, R. and Duncan, W. (1929) On the Criteria for Stability of Small Motions. Proceedings of the Royal Society A, 124, 642-654.

12 Munro, N. (1999) Symbolic Methods in Control System Analysis and Design. IEEE Control Engineering Series, 56, 393 p.

13 Chapellat, H., Battacharyya, S.P and Keel, L.H. (1995) Robust Control: The Parametric Approach. In: Information and System Science Series, N.J. Prentice Hall, Englewood Cliffs.

14 Garloff, J. and Graft, B. (1999) Robust Schur Stability of Polynomials with Polynomial Parameter Dependency. Multidimensional Systems and Signal Processing, 10, 189-199.

15 Jaulin, L. and Walter, E. (1996) Guaranteed Tuning, with Application to Robust Control and Motion Planning. Automatica, 32, 1217-1221.

16 Stoorvogel, A. (1992) The Control Problem. A State Space Approach. Prentice Hall International Series in Systems and Control Engineering, London.

17 Salapaka, M.V., Dahleh, M. and Voulgaris, P. (1997) Mixed Objective Control Synthesis: Optimal l1/h2 Control. SIAM: SIAM Journal on Control and Optimization (SICON), 35, 1672-1689.

18 Zhou, K., Doyle, J.C. and Glover, K. (1996) Robust and Optimal Control. Prentice Hall Inc., Upper Saddle River.

19 Gagnon, E., Pomerleau, A. and Desbiens, A. (1999) Mu-Synthesis of Robust Decentralized PI Controllers. IEEE Proceedings Control Theory Applications, 146, 289-294.

20 Vehí, J. (1998) Analysis and Design of Robust Controllers by Means of Modal Intervals. Ph.D. Thesis, Universitat de Girona, Spain. (in Catalan)

21 SIGLA/X. (1999) Modal Intervals. In: J. Vehí and M. Sainz, Eds., Application of Interval Analysis to Systems and Control, Girona, Spain, 157-221.

22 Fiorio, G., Malan, S., Milanese, M. and Taragna, M. (1993) Robust Performance Design of Fixed Structure Controllers for Systems with Uncertain Parameters. Proceedings of the 32nd Conference on Decision and Control, San Antonio, 15-17 December 1993, 3029-3031.

23 Malan, S., Milanese, M. and Taragna, M. (1997) Robust Analysis and Design of Control Systems Using Interval Arithmetic. Automatica, 33, 1363-1372.

CHAPTER 4

A Robust Multi-RAT VANET/LTE for Mixed Control & Entertainment Traffic

Noha M. Sadek, Hassan H. Halawa, Ramez M. Daoud, Hassanein H. Amer

Electronics and Communications Engineering Department, American University in Cairo, Cairo, Egypt

ABSTRACT

Intelligent Transportation Systems (ITS) have been receiving significant interest from various stakeholders worldwide. ITS promise major enhancements to the efficiency, safety, convenience and sustainability of transportation systems. To satisfy the diverse vehicular application require- ments, this paper proposes an integration of IEEE 802.11-based VANET and LTE cellular network using mobile vehicular gateways. IEEE 802.11 g is used for V2V communications and LTE for V2I communications. A burst communication technique is applied to prevent packet losses in the critical uplink ITS traffic. A performance simulation-based study is conducted to validate the feasibility of the proposed system in an urban vehicular environment. The system performance is evaluated in terms of data loss, data rate, delay and jitter. The results indicate that the proposed Multi-RAT system offers acceptable performance that meets the requirements of the different vehicular applications.

KEYWORDS

Heterogeneous Networks, Intelligent Transportation Systems, Vehicular Networks, VANET, LTE, Wi-Fi

1. INTRODUCTION

Intelligent Transportation Systems (ITS) have been receiving significant interest from various stakeholders worldwide. Automotive companies, governmental entities, standardization organizations and the research community are all focusing on the design and deployment of ITS. ITS promise major enhancements to the efficiency, safety, convenience and sustainability of transportation systems. Specifically, ITS aim at improving road safety, alleviating urban traffic congestion and offering ubiquitous Internet access for passengers [1] .

Vehicular networks are believed to play a crucial role in future ITS due to the wide variety of services that they can offer. Vehicular network applications can be broadly classified into 3 main categories, namely active road safety, traffic efficiency and management, and infotainment applications [2] . Safety applications are the first type of applications that aim at enhancing passengers' safety on the roads by notifying vehicles about dangerous incidents in their vicinity. This can be accomplished by exchanging information like vehicle speed, position and distance heading between vehicles and the infrastructure, which is then used to predict and avoid risky situations [3] .

Besides safety applications, vehicular networks also support traffic efficiency and management applications. The primary objective of applications in this category is to improve traffic flow and enhance traffic coordination and management. Drivers can avoid congestion and find the most efficient route to their destinations with minimum delay.

Additionally, vehicular traffic on the streets will be balanced which will lead to efficient use of the capacity of streets and junctions. This will in turn result in energy savings, pollution reduction and a decrease in travel time. The third type of vehicular applications is infotainment applications, also called comfort or entertainment applications. This class of applications provides information that delivers global Internet services and enhances the passengers' comfort, convenience and entertainment.

In summary, vehicular applications enable the gathering and dissemination of information among vehicles (V2V), and between transportation infrastructure and vehicles (V2I) with the goal of assisting passengers to travel safely, efficiently and conveniently. Reliability, mobility support and low-latency are thus critical to satisfy the performance requirements of the different vehicular applications. On one side of the spectrum, infotainment applications have high-bandwidth demands and QoS-sensitive requirements [4] . While, on the other side of the spectrum, safety-critical applications are characterized by low latency and high message delivery rate. To support safety application demands, a large amount of data traffic needs to be exchanged between vehicles and Base Stations (BSs). This will consequently add a heavy burden to the BSs, which is not likely to be accepted by network operators. Additionally, these extra traffic connections increase the effect of interference and thus increase data error rate. Moreover, this also causes an increase in packet delays due to resource depletion. Furthermore, the scheduler at the BS may have difficulties scheduling transmissions within the tight delay bounds required for safety-critical applications [5] .

To cater to the diverse vehicular application requirements, this paper proposes an integration of IEEE 802.11- based VANET and LTE cellular network using mobile vehicular gateways. The proposed heterogeneous vehicular network combines two technologies with long-range and short-range coverage, namely LTE and Wi-Fi (IEEE 802.11) respectively. Each technology has a different objective and their integrated deployment will improve the vehicular system performance. On one hand, Wi-Fi offers a relatively high capacity at a very low cost and it has a high market penetration. However, it has a low coverage range compared to LTE. This makes it suitable for use as an access network inside the vehicle and for V2V communication between nearby vehicles. On the other hand, LTE offers a wide coverage and better Quality of

Service (QoS) but, it requires costly licensed spectrum and is lacking behind Wi-Fi in terms of economies of scale [6] . These characteristics fit with the long range communication requirements of the V2I backhaul network. By integrating Wi-Fi with LTE, high capacity is coupled with long-range communication to improve the overall performance of the vehicular system. IEEE 802.11 g (Wi-Fi) has been selected because of its popularity and low cost. It offers users a uniform and mass-standard connectivity as the same standard is widely used in various parts of cities like hot-spots, tourist centers and information points [7] .

In the proposed heterogeneous network, there are mainly two types of vehicles: Gateway Vehicles (GVs) and Ordinary Vehicles (OVs). GVs are equipped with both LTE and Wi-Fi interfaces while OVs are only Wi-Fi enabled. A GV can be connected to two networks simultaneously: the LTE network using its E-UTRAN interface and to other OVs through its Wi-Fi interface. The GV can thus serve as a mobile gateway (i.e., relay node) for other OVs in its vicinity to access the LTE network. This can be accomplished by receiving data from OVs (using the Wi-Fi interface) and relaying it to the LTE network (via the LTE interface). OPNET Modeler [8] is used for the design, implementation and evaluation of the proposed network model.

The rest of the paper is organized as follows. Section 2 provides a background about vehicular networking research. Section 3 describes the proposed architecture. Section 4 presents the results and performance evaluation of the proposed architecture. Finally, the paper is concluded in Section 5.

2. BACKGROUND

This section presents some studies that evaluated the feasibility of using IEEE 802.11 and cellular networks to support vehicular networking applications.

In [9], an attempt was made to solve the traffic control problem in an urban environment using Wi-Fi where a burst communication technique was used to eliminate packet losses. The authors in [10] proposed a heterogeneous LTE/Wi-Fi vehicular system that supports both infotainment and ITS traffic control data. Other studies ([1] [5] [11] - [14]) evaluated the performance of heterogeneous LTE-VANET (Vehicle Ad-hoc Networks) networks, where LTE was the backhaul network and VANET was used for inter-vehicle communication. In [5], LTE was used to exchange Cooperative Awareness Messages (CAMs) between clusters while, Wi-Fi delivered in-cluster information. Reference [11] envisioned a heterogeneous LTE/IEEE 802.11p network that provides multimedia communication services over spatially apart vehicular groups and a cluster head election mechanism was proposed. A cooperative protocol based on coalition game theory was introduced in [12] to disseminate data in LTE/VANET network. Reference [13] investigated the delivery of real-time streaming of scalable video coded (SVC) video over vehicle-to-infrastructure (V2I) links. The inter-vehicles to infrastructure (V2V2I) model considered in [1] again used IEEE 802.11p for V2V communications and LTE for V2I communications. In [14] , the LTE network was used as a cluster management infrastructure for the IEEE 802.11p VANET. The performance was compared with the decentralized VANET architecture for an urban sensing application. Many of the studies in the field of vehicular communications only investigate a single type of application and either cover V2I or V2V communication only. The simultaneous support of safety, traffic and infotainment applications using both V2I and V2V communication has not been addressed in the literature. Accordingly, this perspective will be investigated in this study.

3. PROPOSED MODEL

3.1. Network Architecture

The envisioned WiFi-based VANET-LTE heterogeneous network architecture is shown in Figure 1. Vehicles that are equipped with both LTE and Wi-Fi interfaces are referred to as Gateway Vehicles (GVs) whereas, only Wi-Fi is supported on-board Ordinary Vehicles (OVs). A GV is under the coverage region of at least one LTE eNodeB, and its LTE and Wi-Fi interfaces are both activated. On the other hand, an OV either lacks an LTE interface or is not present in an LTE coverage area.

In other words, it is assumed that the LTE interface is either absent or disabled on OVs.

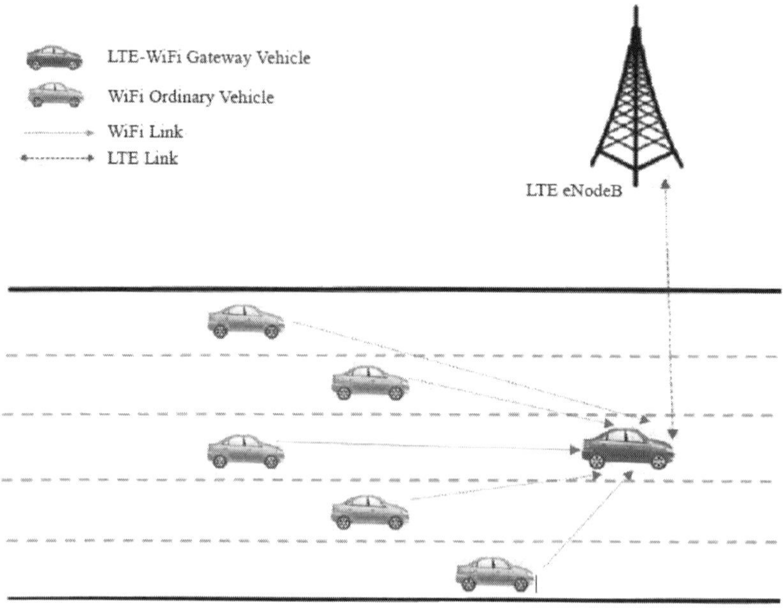

Figure 1. VANET-LTE network architecture.

This paper builds on the model introduced in [10] which consists of seven cells arranged in a hexagonal honey-cell layout where each cell is covered by one eNodeB. The vehicles are modeled moving in a radial path between the 7 cells in an urban traffic model where vehicles move at the maximum speed of 60 Km/hr. The proposed architecture divides the vehicular network into Vehicle-to-Vehicle (V2V), Vehicle-to-Infrastructure (V2I), On-board Vehicle Communication (OVC), and backhaul connection. The V2V network allows communication between GVs and OVs through Wi-Fi. The V2I network between the vehicle and LTE eNodeB provides access to the LTE core network. The OVC network consists of a Wi-Fi Access Point (AP), passengers' devices and a vehicular OnBoard Unit (OBU) for ITS data. In this scenario, LTE is the backhaul link used to access the Internet and the connectivity is shared to vehicular users using Wi-Fi as the last mile link. The LTE backhaul links are full duplex. Typically, an LTE eNodeB is deployed alongside the road and the vehicles are under the coverage of the different eNodeBs. More details about the system architecture can be found in [10] .

From the data flow perspective, GVs sample and gather the information from OVs (through Wi-Fi) then, in turn send periodically the relevant data to the infrastructure (via LTE). Data is exchanged between the GVs and LTE eNodeBs in both the downlink and uplink directions. On the downlink, LTE eNodeBs unicast the data to the GVs where both ITS and infotainment traffic are sent over the LTE link yet, they are routed differently inside the vehicle based on the intended destination. ITS data is sent to the vehicle's OBU via Ethernet for further processing and decision making. On the other hand, infotainment traffic is sent to the passengers' devices through Wi-Fi. On the uplink, GVs forward the traffic data (collected from OVs) to the LTE infrastructure at a pre-determined transmission rate.

In order to reduce the amount of traffic exchanged between vehicles and eNodeBs, a clustering strategy is employed. The GV is the cluster head which maintains the status of the cluster. Only the LTE-enabled cluster head is allowed to receive/transmit data from/to eNodeBs through LTE interfaces. Every OV transmits small data packets called Cooperative Awareness Messages (CAMs) to their corresponding cluster head/GV, providing state information such as speed or location. Such CAMs are transmitted every 100 ms [5] . It is considered that the vehicle clusters are already formed and that in each cluster, vehicles are moving together across all seven cells along the trajectory described in [10] . Thus, cluster members can be assumed not to vary throughout the journey. A single cluster of vehicles consists of 1 GV and 5 OVs.

3.2. Simulation Parameters

The network model is simulated using OPNET Modeler [8] network simulator. The various network configuration parameters are shown in Table 1. eNodeB, GV LTE and downlink ITS parameters are all obtained from the model explained in [10] . CAMs are transmitted from OVs to GV where CAM size is 40 Bytes and transmission interval is 100 ms [5] . GV then aggregates CAMs' data (12000 Bytes) and sends it on the LTE uplink channel every 30 seconds. 33 simulation runs were performed with different random seeds in order to ensure statistical accuracy. It is important to note that all results presented in this paper are subjected to a 95% confidence analysis.

3.3. Performance Evaluation Metrics

This section describes the performance evaluation metrics of the proposed model. The network performance is evaluated in terms of data rate, Data Loss Ratio (DLR), delay and jitter parameters, defined as follows:

- Data Rate (in Bytes/sec) is defined as the sum of the data bytes received at the destination averaged over time.
- Data Loss Ratio (DLR) is defined as the ratio between dropped packets that do not reach the destination and the total number of packets sent from the source to the destination.
- Delay (in seconds) specifies the time elapsed between sending the request from the source and the reception of the response at the source. This metric serves as a measure of the average overall delay of the packets for a particular node.
- Jitter (in seconds) is defined as the packet delay variation. This metric is calculated as the standard deviation of packet delay for all packets sent over the network for a particular node.

4. SIMULATION RESULTS

In this paper, the overall system performance, as specified by the communication requirements imposed by different types of vehicular networking applications, is investigated. For this purpose, the foremost emphasis is on evaluating the data loss, data rate, delay and jitter of vehicular V2V and V2I applications in a realistic urban simulation environment. For all presented scenarios, the aforementioned performance evaluation metrics (data rate, DLR, delay and jitter) were analyzed. A 95% confidence analysis was performed for all presented results.

Table 1. Simulation parameters.

Parameter	Value
eNodeB	
Transmit Power	10 Watts
Antenna Gain	18 dBi
MIMO	2 × 2
Bandwidth	10 MHz
Frequency Band	1.8 GHz
Rx Sensitivity	–123 dBi
Duplexing Technique	FDD
Cell Radius	1.5 Km
ISD	2.6 Km
Transmit Power	10 Watts
GV	
Transmit Power	0.2 Watts
Antenna Gain	0 dBi
MIMO	1 × 2
Rx Sensitivity	–106 dBi
Shadow Fading Standard Deviation	4 dB
Downlink ITS IPT	120 s
Downlink ITS Size	1024 Bytes
Uplink ITS IPT	30 s
Uplink ITS Size	12000 Bytes
OV	
CAM Transmission Interval	100 ms
CAM Size	40 Bytes

4.1. No Burst Model

In a video streaming service environment, it is important to maintain the DLR threshold below 1% [15] [16] such that the QoS requirement of video streaming service users is satisfied. Additionally, the performance of video streaming depends greatly on delay and jitter. According to [17] , the maximum acceptable video packet delay is set to 150 ms and the maximum allowable jitter is 50 ms. For traffic control data, since most of the applications are time-critical, the end-to-end delay must be between 100 and 500 ms [18] .

Table 2 shows the results obtained for vehicular video, downlink and uplink ITS traffic. It can be seen that the maximum video streaming traffic delay is 12.94 ms, while the maximum jitter is 5.84 ms. For downlink ITS traffic, a maximum delay of 22.27 ms and jitter of 6.7 ms is observed. As for uplink ITS traffic, the maximum delay is 23.82 ms and the maximum jitter is 9.69 ms. The obtained values are all below the above-mentioned benchmarks for ITS applications. It can be noted that the DLR for uplink traffic is 1.73%.

Figure 2 shows two packet losses in the uplink data traffic; one of them happens when the GV is in LTE cell 1 while the other one is in LTE cell 2. Due to the criticality of the safety-related information communicated in the uplink direction, a zero DLR is desired. So, in the next section, a burst recovery technique will be proposed to mitigate data losses in uplink ITS traffic.

The Wi-Fi V2V traffic results are presented in Table 3. Worst case values are considered where the DLR, delay and jitter values represent the upper bound of the resulting confidence interval. No data drops were reported for any of the OVs. In other words, all V2V data was successfully received by the GV via Wi-Fi. Additionally, the obtained delay values are far below the 100 ms constraint of the CAM V2V communication.

4.2. Burst Recovery Mechanism

Burst is a communication technique used to reduce or prevent data losses in wireless communication systems. This is accomplished by sending successive identical packets within a certain time frame and separated by a pre-defined period of time. In essence, if one of the original packets is lost, the other redundant packets will still carry the same information to the desired destination.

There are basically 3 parameters that characterize burst communication, namely T_{packet}, T_{F2L} and T_{burst} [9] . T_{packet} is the time between two successive packets within the same burst. T_{F2L} is the time between the first and the last packet in the same burst. Finally, T_{burst} is the time between the first packets of two successive bursts.

T_{packet} depends on the Inter-Packet Transmission Time (IPT) of the uplink traffic and on number of packets used in one burst, as follows:

Table 2. Confidence results of video, downlink, and uplink ITS traffic—No burst model.

Metric	Video	Downlink ITS	Uplink ITS
Data Rate (Bytes/sec)	[124311.1; 124338.5]	[8.46; 8.56]	[393.07; 397.69]
DLR (%)	0.55%	0.8%	1.73%
Delay (ms)	[12.91; 12.94]	[20.97; 22.27]	[22.79; 23.82]
Jitter (ms)	[5.59; 5.84]	[4.48; 6.7]	[7.47; 9.69]

Table 3. Results of V2V traffic—No burst model.

Metric	V2V OV1	V2V OV2	V2V OV3	V2V OV4	V2V OV5
Data Rate (Bytes/sec)	400	400	400	400	400
DLR (%)	0%	0%	0%	0%	0%
Delay (ms)	0.790889	1.11316	1.407288	1.63815	0.4702
Jitter (ms)	0.443402	0.41698	0.387431	0.35542	0.41207

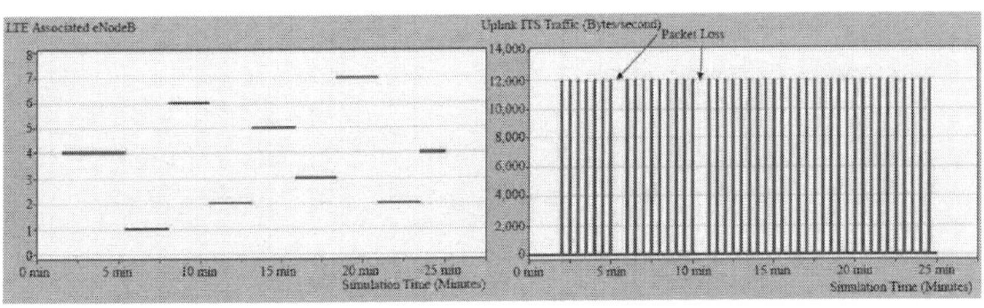

Figure 2. Uplink ITS traffic and LTE associated eNodeB.

$$T_{\text{packet}} \leq \frac{IPT}{N_{\text{burst}}} \tag{1}$$

where IPT is the Inter-Packet Transmission Time of the uplink traffic and N_{burst} is the number of packets per burst. T_{F2L} depends on the number of packets used in one burst and must satisfy the following constraint:

$$T_{F2L} = T_{\text{packet}} \times \left(N_{\text{burst}} - 1\right) \tag{2}$$

T_{burst} must be less than or equal the IPT, which corresponds to the time needed for traffic and routing updates.

$$T_{\text{burst}} \leq IPT \tag{3}$$

The minimum number of packets that can be used in one burst is two. A two-packet burst is studied to optimize the LTE channel utilization and minimize the network load. From Equation (1) and for a 30-sec uplink IPT, Tpacket must be smaller than or equal to 15 sec. In case of 2 packets per burst, $T_{F2L} = T_{packet}$ from Equation (2).

Figure 3 shows a typical burst communication for ITS uplink traffic. It is clear that two packets were lost from the original uplink traffic, however their burst replicas arrived successfully which indicates that no actual uplink ITS data was lost. Table 4 shows the results obtained for vehicular video, downlink and uplink ITS traffic for the Burst model. Table 5 shows Wi-Fi V2V traffic results for the Burst model. All received data, delay and jitter are within the acceptable limits of ITS applications.

To conclude, the delay, jitter and data drops for both uplink and downlink ITS traffic, and video traffic are within the acceptable limits using the burst technique with only two packets per burst. The obtained simulation results thus show that the proposed system simultaneously satisfies vehicular and infotainment application requirements.

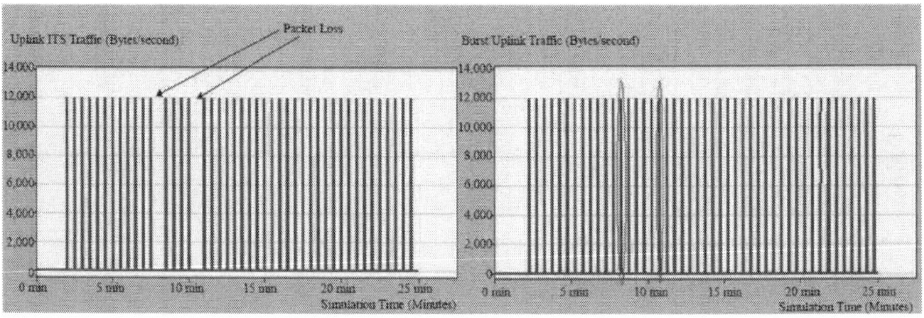

Figure 3. Uplink ITS traffic with Burst.

Table 4. Confidence results of video, downlink, and uplink ITS traffic—Burst model.

Metric	Video	Downlink ITS	Uplink ITS
Data Rate (Bytes/sec)	[124321.4; 124347.2]	[8.39; 8.54]	[393.08; 397.1]
DLR (%)	0.5%	1.8%	0%
Delay (ms)	[12.9; 12.93]	[21.52; 22.53]	[23.4; 24.36]
Jitter (ms)	[5.56; 5.66]	[5.32; 7.58]	[8.69; 11.13]

Table 5. Results of V2V traffic—Burst model.

Metric	V2V OV1	V2V OV2	V2V OV3	V2V OV4	V2V OV5
Data Rate (Bytes/sec)	400	400	400	400	400
DLR (%)	0%	0%	0%	0%	0%
Delay (ms)	0.79	1.113	1.406	1.637	0.47
Jitter (ms)	0.4436	0.4178	0.3867	0.3547	0.4119

5. CONCLUSIONS

It is anticipated that ITS will play a vital future role in improving traffic efficiency, safety, comfort and emissions. Such systems rely on advanced mobile and wireless communication systems to satisfy the different vehicular applications' requirements and to accommodate the increasing vehicles' fleet. In this paper, an integrated IEEE802.11g-based VANET and LTE heterogeneous vehicular network was proposed where infotainment traffic was sent in addition to ITS control traffic in an urban vehicular environment. The system performance was evaluated in terms of data loss ratio, data rate, delay and jitter.

Data losses in uplink ITS data traffic were initially observed so, a Burst technique was proposed to prevent packet losses. A quantitative analysis was performed to determine the number of packets per burst, the inter-packet and inter-burst intervals. It was found that a substantial improvement was achieved using a two-packet Burst, where no packets were lost in the uplink direction. Additionally, for the given simulation scenario and network traffic load, it was shown that the proposed system meets both the video and ITS traffic application requirements. Thus, the feasibility of the proposed VANET-LTE heterogeneous system in urban vehicular environments was verified. Future work will focus on investigating more applications in the area of ITS.

REFERENCES

1 Imadali, S., Kaiser, A., Decremps, S., Petrescu, A. and Vèque, V. (2013) V2V2I: Extended Inter-Vehicles to Infrastructure Communication Paradigm. IEEE 5th Global Information Infrastructure Symposium, Trento, Italy, October 2013.

2 Hossain, E., et al. (2010) Vehicular Telematics over Heterogeneous Wireless Networks: A Survey. Computer Communications, 33, 775-793.

3 Karagiannis, G., et al. (2011) Vehicular Networking: A Survey and Tutorial on Requirements, Architectures, Challenges, Standards and Solutions. IEEE Communications Surveys & Tutorials, 13, 584-616.

4 Araniti, G., Campolo, C., Condoluci, M., Iera, A. and Molinaro, A. (2013) LTE for Vehicular Networking: A Survey. IEEE Communications Magazine, 51, 148-157.

5 Tung, L., Mena, J., Gerla, M. and Sommer, C. (2013) A Cluster Based Architecture for Intersection Collision Avoidance Using

Heterogeneous Networks. 12th IFIP/IEEE Annual Mediterranean Ad Hoc Networking Workshop (Med- Hoc-Net), Cesson-Sevigne, France, June 2013.

6 Ghaleb, A., et al. (2013) Preservation of QOS across Hybrid LTE-WLAN Router. IET International Conference on Information and Communications Technologies (IETICT), Beijing, April 2013.

7 Barbu, G. (2010) E-TRAIN-Broadband Communication with Moving Trains, Technical Report—Technology State of the Art. International Union of Railways, Paris.

8 Official Site of OPNET: www.opnet.com

9 Daoud, R., El-Dakroury, M., Elsayed, H., Amer, H. and ElSoudani, M. (2007) WiFi Architecture for Traffic Control Using MIPv6. IEEE Mediterranean Conference on Control and Automation MED, Athens, 27-29 July 2007, 1-5.

10 Sadek, N., Halawa, H., Daoud, R., Amer, H. and Ali, N. (2015) Heterogeneous LTE/Wi-Fi Architecture for ITS Traffic Control and Infotainment. International Conference on Electrical Systems for Aircraft, Railway, Ship Propulsion and Road Vehicles, Aachen, March 2015.

11 Sivaraj, R., Gopalakrishna, A., Chandra, M. and Balamuralidhar, P. (2011) QoS-Enabled Group Communication in Integrated VANET-LTE Heterogeneous Wireless Networks. IEEE 7th International Conference on Wireless and Mobile Computing, Networking and Communications (WiMob), October 2011.

12 Li, Y., Ying, K., Cheng, P., Yu, H. and Luo, H. (2012) Cooperative Data Dissemination in Cellular-VANET Heterogeneous Wireless Networks. 4th International High Speed Intelligent Communication Forum (HSIC), Nanjing, May 2012.

13 Yaacoub, E., Filali, F. and Abu-Dayya, A. (2013) SVC Video Streaming over Cooperative LTE/802.11p Vehicle-to- Infrastructure Communications. IEEE World Congress on Computer and Information Technology (WCCIT), Tunis, June 2013.

14 Remy, G., Senouci, S., Jan, F. and Gourhant, Y. (2011) LTE4V2X: LTE for a Centralized VANET Organization. IEEE Global Telecommunications Conference (GLOBECOM), Houston, 5-9 December 2011, 1-6.

15 Islam, M. and Chowdhury, M. (2013) Study of Inter-Cell Interference and Its Impact on the Quality of Video Conference Traffic in LTE Network. M.Sc. Thesis, Blekinge Institute of Technology, Sweden.

16 Ramli, H., Basukala, R., Sandrasegaran, K. and Patachaianand, R. (2009) Performance of Well Known Packet Scheduling Algorithms in The Downlink 3GPP LTE system. IEEE 9th Malaysia International Conference on Communications (MICC), Kuala Lumpur, 15-17 December 2009, 815-820.

17 International Telecommunication Union (ITU) (2008) Subjective Video Quality Assessment Methods for Multimedia Applications. Recommendation ITU-T P.910, Geneva.

18 Mir, Z. and Filali, F. (2014) LTE and IEEE 802.11p for Vehicular Networking: A Performance Evaluation. EURASIP Journal on Wireless Communications and Networking, 2014, 89.

CHAPTER 5

Experimental Investigation on Adaptive Robust Controller Designs Applied to Constrained Manipulators

Samuel L. Nogueira [1,*], Tatiana F. P. A. T. Pazelli [2], Adriano A. G. Siqueira [1] and Marco H. Terra [3]

[1] Department of Mechanical Engineering, University of São Paulo, 400 Trabalhador São-Carlense av., 13566-590, São Carlos, Brazil

[2] Department of Electrical Engineering, Federal University of São Carlos,Washington Luís, km 235-SP-310, São Carlos, Brazil

[3] Department of Electrical Engineering, University of São Paulo, 400 Trabalhador São-Carlense av., 13566-590, São Carlos, Brazil

ABSTRACT

In this paper, two interlaced studies are presented. The first is directed to the design and construction of a dynamic 3D force/moment sensor. The device is applied to provide a feedback signal of forces and moments exerted by the robotic end-effector. This development has become an alternative solution to the existing multi-axis load cell based on static force and moment sensors. The second one shows an experimental investigation on the performance of four different adaptive nonlinear H_∞ control methods applied to a constrained manipulator subject to uncertainties in the model and external disturbances. Coordinated position and force control is evaluated. Adaptive procedures are based on neural networks and fuzzy systems applied in two different modeling strategies. The first modeling strategy requires a well-known nominal model for the robot, so that the intelligent systems are applied only to estimate the effects of uncertainties, unmodeled dynamics and external disturbances. The second strategy considers that the robot model is completely unknown and, therefore, intelligent systems are used to estimate these dynamics. A comparative study is conducted based on experimental implementations performed with an actual planar manipulator and with the dynamic force sensor developed for this purpose.

KEYWORDS

constrained manipulators; $\mathcal{H}_\infty$control; neural networks; fuzzy systems; variable structure control; load cell; multi-axis force sensor

1. INTRODUCTION

The first robotic manipulators were developed in order to perform positioning tasks. They were then specifically designed to be robust enough so as to not be affected by external disturbances. This physical robustness of robot manipulators has enabled researchers to obtain accurate positioning systems based on simple control laws. Decades later, the popularization of industrial robotics has heightened researchers' interest in creating a much wider range of applications for robotic manipulators in various environments.

Nowadays, many applications demand robotic manipulators to perform tasks subject to force and motion constraints. For example, the process of milling a piece requires accurate incidence angles, paths, forces and moments exerted by the drill in the milled material. Additionally, in industrial assembly lines, the objects must be assembled along certain paths with predetermined forces and moments. In sheet metal cutting, the cutting angles, paths and forces exerted on the material are also important. Moreover, on surfaces where polishing disks must always be perpendicular to the surface being polished, predetermined force must be applied. Consequently, new concepts of position and force control for lighter and more flexible robots have been created [1,2].

The problem defined in these applications involves three stages: the approach phase, the impact moment and the sustained contact tracking. The approach phase has been addressed in many works and defines the problem of positioning the tool without, or before, touching the environment. The second phase requires controlling the initial impact and damping out the vibrations generated during the event. After the initial impact, sustained contact is desired in many operations. In these cases, not only the motion of the end-effector is required to follow a prescribed path, but also the force exerted by the end-effector is required to follow a predefined reference. In these constrained systems, forces and moments generated between the end-effector and the target must be controlled, rather than being treated as disturbances and rejected. Addressing manipulators subject to model uncertainties and disturbances, the work considered in this paper is concentrated on the sustained contact tracking phase of the problem.

The great majority of solutions presented in the literature for the problem of constrained systems control require knowing the forces and moments of interaction between the robot end-effector and the environment where it is acting. The problem of force control undertaken in this paper was also considered in [1,3–7]. However, only simulation results were presented.

As a differential and important contribution, this paper addresses an experimental investigation on robust force control as a result of the development of a modular sensor device. The proposed device is designed and built to measure dynamic forces and moments in three orthogonal axes based on unidirectional force sensor units. As a consequence of its independent architecture on the type of sensitive material, static or dynamic force sensors can be applied. Piezoelectric or piezoresistive force sensors are effective solutions for the applications involved in this study due to their inherent dynamic response characteristics.

This paper is organized as follows: next section presents preliminary concepts and relevant results found in the literature; Section 3 introduces the model description of the constrained robot manipulator; Section 4 presents the problem formulation; Section 5 describes the solutions for the nonlinear H_∞ control problems based on the linear parametrization property of the model, neural networks and fuzzy systems; Section 6 demonstrates the 3D dynamic force and moment sensor; and Section 7 presents the experimental results for a three-link manipulator.

2. PRELIMINARY CONCEPTS

The concept of stiffness control was introduced by Salisbury [8]. It is based on the resistance of the environment in which the robotic end-effector applies the force. The problem is modeled as a mass-spring system and this method made possible the simultaneous position/force control. However, it considers constant desired position and force. In many robotic applications, such as when milling a piece, the end-effector must follow a trajectory along the surface of an object while applying a desired force, which is not necessarily constant. In this case, the stiffness control application does not work properly.

To address this type of limitation, Raibert and Craig [9] partitioned the control problem into two subtasks: one task is for controlling the position trajectory and the other task for controlling the desired force. This approach has been evolutionary for controllers, as proposed by Paul et al. [10], and became the conceptual basis of the hybrid trajectory of position and force control currently found in the literature.

It was shown by McClamroch and Wang [11], that when a manipulator is in contact with a surface, the position degrees of freedom are reduced. In this case, force constraint is added to motion equations through Lagrange multipliers. Thus, the order of the state vector is reduced in the dynamic equations of the manipulator.

Since the phase of controlling robotic systems with restricted position and force has been overcome, researchers began to focus on variables that could degrade the stability of the proposed models. The load on the end-effector that can fluctuate while the manipulator performs various tasks, friction and parameter uncertainties, are some examples that have required much research effort [12–15]. However, only a few studies have dealt with adaptive and robust control of robotic systems subject to constraints [16–20]. Using the basis developed by [9,11], the work performed by Chang and Chen [21] submitted an adaptive controller with H_∞ robust performance criterion for robotic systems with position and force constraints.

On the development of adaptive robust controllers, neural networks, combined with a nonlinear H_∞ control law, were applied by Chang and Chen [22] to control robotic systems subject to parametric uncertainties and external disturbances. A smooth response was achieved with a simple and computationally efficient implementation. Neural networks were applied to estimate the unknown dynamics of the system, thereby not requiring mathematical modeling knowledge. A variable structure controller (VSC) is added to this formulation by Chang and Chen [23]. The inclusion of VSC in the control law weakens the hypothesis used by the authors in [22] that the estimation error should be integrable, and limits it to be only a state-dependent function. Following this approach, the authors in [24] developed an adaptive H_∞ controller based on fuzzy systems and VSC for robots with position and force constraints.

Karayiannidis and Doulgeri proposed adaptive controllers in [5,6] for force and position trajectory tracking in environments with little or no constraint knowledge. In these papers, the main purpose is to explore the workspace using measurements of position, speed and force in a robotic end-effector, to attenuate impacts caused by unknown environments. In order to improve estimates of these models, the use of a camera was proposed by Cheah et al. [4] to provide a better estimate of the contact constraint through computer vision. However, these works are more interested in exploring the environmental constraint, rather than a repetitive task control with great tracking precision.

To satisfy the need of measuring the forces and moments of interaction between the robot end-effector and the environment, devices known as multi-axis load cell or multi-axis force and moment sensor are used. There are several patents on devices whose purpose is to measure forces and moments in three axes.

Force sensors, such as strain gages, have been generally used as the basic unit of measurement, as shown by the illustrative examples in Figure 1. The multi-axis load cell developed by Meyer and Lowe [25] was built in one piece with internal and external parts that are connected by a pair of axially spaced beams. The beams are fixed in the center of the piece, where the strain gages are attached, and an outside tunnel. The loads are measured by the curvature of the connectors. The load transducer developed by Meyer et al. [26] measures linear forces in three axes and moments of about two axes. The transducer has encapsulations connected by internal and external arms-sensitive loading. The device described by Sommerfeld et al. [27] consists of an external annular body, a hub and four beams that hold the hub to the radially outward portion. Strain gages are fixed to the faces of the beams, with a 90 degree lagging. Forces and moments exerted on the hub are transmitted to the four beams, and consequently, to the strain gages. Differently from [25,26], this device described by Sommerfeld et al. [27] is able to measure forces and moments in three orthogonal axes. However, all the devices mentioned above have an extrusion completely dependent upon the type of sensitive material being used.

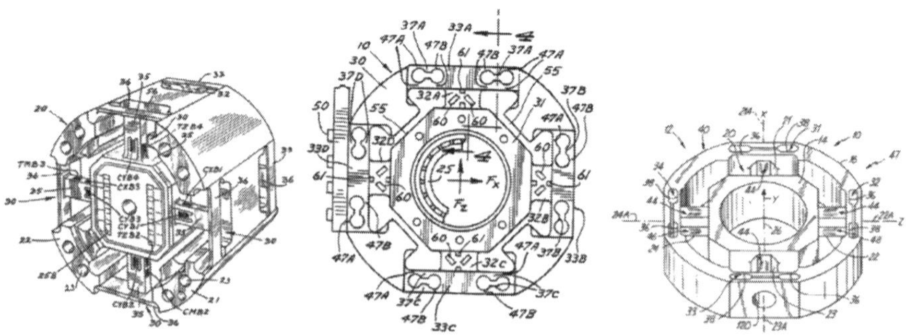

(a) (b) (c)

Figure 1. (**a**) Image taken from [25]; (**b**) Image taken from [26]; (**c**) Image taken from [27].

3. MODEL DESCRIPTION

Let a constrained robot manipulator be defined by an n-link serial-chain rigid manipulator whose end-effector is in contact with a rigid environmental constraint, according to Figure 2.

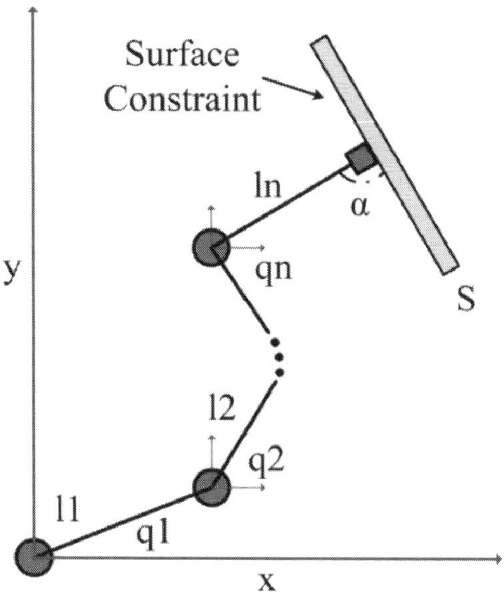

Figure 2. Constrained robot manipulator.

The links of the manipulator are numbered from 1 to n: J_i is the joint connecting the (i − 1)-th and i-th links, l_i is the vector connecting J_i and J_{i+1}, q_i is the angle of the i-th link around joint J_i, C_i is the center of mass of the i-th link and l_{ci} is the vector connecting J_i and C_i. The constraint surface is represented by S and α is the angle of contact between the end-effector and S.

Remark 1 *Assume that the end-effector is already in contact with the constraint surface, and the control exerted over the constraint force is such that the force will always maintain the end-effector in contact with the constraint surface.*

3.1. Robot Dynamics

The dynamic equations of a constrained robot is given from Lagrange theory as

$$M(q)\ddot{q} + C(q,\dot{q})\dot{q} + G(q) = \tau + f + \tau_d \tag{1}$$

where $M(q) \in \Re^{n\times n}$ is the symmetric positive definite inertia matrix, $C(q,\dot{q}) \in \Re^{n\times n}$ is the Coriolis and centripetal matrix, $G(q) \in \Re^{n}$ is the vector of the gravitational torques, $\tau \in \Re^{n}$ is the torque vector acting upon the manipulator joint, $f \in \Re^{n}$ denotes the vector of joint-space generalized forces on the environmental constraint exerted by the end-effector, and τ_d defines finite energy unknown disturbances.

Model uncertainties in Equation (1) can be introduced dividing the matrices M(q), C(q, q˙), G(q), and f into a nominal and a perturbed part:

$$\begin{aligned} M(q) &= M_0(q) + \Delta M(q) \\ C(q,\dot{q}) &= C_0(q,\dot{q}) + \Delta C(q,\dot{q}) \\ G(q) &= G_0(q) + \Delta G(q) \\ f &= f_0 + \Delta f \end{aligned} \tag{2}$$

where $M_0(q), C_0(q,\dot{q}), G_0(q)$, and f_0 are nominal matrices and $\Delta M(q), \Delta C(q,\dot{q}), \Delta G(q)$, and Δf represent the uncertainties.

3.2. Constraint Modelling

Considering **Remark 1**, the m-dimensional surface constraint is described by the holonomic relationship

$$\phi(q) = 0 \tag{3}$$

where $\phi(q) : \Re^n \to \Re^m$ is a smooth function.

Constraint forces are given by

$$f = J_c^T(q)\lambda \tag{4}$$

where $J_c(q) = \frac{\delta\phi(q)}{\delta q} \in \Re^{m\times n}$ is the Jacobian matrix that relates the constraint to the controlled variables of the robot and $\lambda \in \Re^m$ is a vector of generalized Lagrangian multipliers associated with the constraint.

In this paper, it is considered that parametric uncertainties may also be included into the constraint model since the constraint surface may be not perfectly rigid, frictionless or even that its geometric description may not be exactly known. Thus, consider

$$\begin{aligned} \phi(q) &= \Delta\phi(q) \\ J_c(q) &= J_c(q) + \Delta J_c(q) \end{aligned} \tag{5}$$

and assume that $\Delta\phi$ and ΔJ_c are implicit in Δf, described in Equation (2).

3.3. Reduced Order Model

The presence of m constraints causes the manipulator to lose m degrees of freedom, and therefore, n – m linearly independent coordinates are sufficient to characterize the constrained movement. Therefore, to formulate a reduced order dynamics for the constrained system, the formulation presented in this paper follows the assumptions made by McClamroch and Wang [11] and posteriorly by Chang and Chen [24].

Define

$$q = \begin{bmatrix} q^1 \\ q^2 \end{bmatrix}$$

where $q^2 = \sigma(q^1)$ and $\phi(q^1, \sigma(q^1)) = 0$.

The following reduced model formulation is obtained for the constrained manipulator as the model proposed by Chang and Chen [24]:

$$A_L(q^1)\ddot{q}^1 + L^T(q^1)C_L(q^1, \dot{q}^1)\dot{q}^1 + L^T(q^1)G(q^1) = L^T(q^1)(\tau + \tau_d) \tag{6}$$

where

$$L(q^1) = \begin{bmatrix} I_{(n-m)} \\ \frac{\partial \sigma(q^1)}{\partial q^1} \end{bmatrix}$$

and

$$C_L(q^1, \dot{q}^1)\dot{q}^1 = M(q^1)\dot{L}(q^1) + C(q^1, \dot{q}^1)L(q^1)$$

The necessary structure and properties of the model for controller formulation are maintained since $A_L(q^1) = L^T(q^1)M(q^1)L(q^1)$ is symmetric positive definite and the matrix $\dot{A}_L(q^1) - 2L^T(q^1)C_L(q^1, \dot{q}^1)$ is skew symmetric.

4. PROBLEM FORMULATION

Let $q_d(t) \in \Re^n$ and $\dot{q}_d(t) \in \Re^n$ be the desired reference trajectory and the corresponding velocity for the joints, respectively Assume that $q_d(t)$ and its derivatives $\dot{q}_d(t)$ and $\ddot{q}_d(t)$ are bounded. Define a bounded $f_d \in \Re^n$ as the desired reference contact force. To be consistent with the imposed restrictions, we need to assure that $\phi(q_d) = 0$ and $f_d = J_c^T(q_d)\lambda_d$.

Since $q^2 = \sigma(q^1)$, it is only necessary to find a control law that makes $q^1 \to q_d^1$ when t $\to$ ∞. Therefore, define the position tracking error $\bar{x}_1(t)$ and the filtered link tracking error $\bar{x}_1(t)$ as in [24]:

$$\bar{x}(t) \doteq \begin{bmatrix} \bar{x}_1(t) \\ \bar{x}_2(t) \end{bmatrix} \doteq \begin{bmatrix} q^1(t) - q_d^1(t) \\ \dot{q}^1(t) - \dot{q}_d^1(t) + p(q^1(t) - q_d^1(t)) \end{bmatrix} \tag{7}$$

for some constant p > 0.

From Equations (6) and (7), the error dynamic equations can be obtained as

$$\dot{\bar{x}} = \begin{bmatrix} \dot{\bar{x}}_1 \\ \dot{\bar{x}}_2 \end{bmatrix} = A\bar{x} + Bu + B\omega \tag{8}$$

where

$$A = \begin{bmatrix} -pI & I \\ 0 & -A_L^{-1}(q^1)C_L(q^1,\dot{q}^1) \end{bmatrix}, \quad B = \begin{bmatrix} 0 \\ -(M(q^1)L(q^1))^{-1} \end{bmatrix}$$

$$u = F(x_e) - \tau, \quad \omega = \tau_d, \quad \text{and}$$

$$F(x_e) \doteq M(q^1)L(q^1)(\ddot{q}_d^1 - p\dot{\tilde{x}}_1) + C_L(q^1,\dot{q}^1)(\dot{q}_d^1 - p\tilde{x}_1) + G(q^1)$$

Within this problem formulation, the torques applied to the joints to guarantee the task execution are given by

$$\tau = F(x_e) - u \tag{9}$$

where the term $F(x_e)$ refers to the dynamics of the controlled variables and u is the control law provided by the adaptive $\mathcal{H}_\infty$ controller proposed by Chang and Chen [24].

The term $F(x_e)$ can be divided in a nominal and uncertainty part, that is,

$$F(x_e) = F_0(x_e) + \Delta F(x_e)$$

where term $\Delta F(x_e)$ contains the uncertainties from parametric Equations (2) and (5).

In this paper, the control problem is solved based on two different approaches. In the first approach, an adaptive intelligent system is applied to estimate only the term $\Delta F(x_e)$, considering that the nominal model of the robot is well known. In the second approach, it is considered that the system model or the term $F_0(x_e) + \Delta F(x_e)$ is completely unknown, then the adaptive intelligent system is applied to estimate it. The nonlinear H_∞ control and VSC are applied in both approaches to attenuate the effects of estimation errors and external disturbances.

5. ADAPTIVE NONLINEAR H_∞ CONTROLLER BASED ON INTELLIGENT SYSTEMS

The adaptive control laws presented as follows are based on two different learning methods to estimate uncertain parameters and also the behavior of unmodeled dynamics. Neural networks and fuzzy systems based on the Takagi–Sugeno model are considered. The structure of these intelligent systems can be seen in details in [28].

Let $\hat{F}(x_e, \Theta) = \Xi(x_e)\Theta$ be the output of the adaptive intelligent system, where x_e is the input vector and Θ is a vector of adjustable parameters such that $\Theta^T\Theta \leq M_\theta | M_\theta > 0$. Two different cases are considered in the following:

Case 1 *Estimation of Model Uncertainties Based on Intelligent Systems*

In this case, the nominal model of the robot is considered as well known, the intelligent system then estimates only the model uncertainties, such as

$$\hat{F}_1(x_e, \Theta) \approx \Delta F(x_e)$$

Therefore u and ω in Equation (8) will be rewritten as

$$u = F_0(x_e) + \hat{F}_1(x_e, \Theta) - \tau$$

$$\omega = (\Delta F(x_e) - \hat{F}_1(x_e, \Theta)) + \tau_d$$

so that ω includes the estimation error from the intelligent system.

Case 2 *Estimation of Complete Model Based on Intelligent Systems*

In this case, however, the nominal model of the robot is considered completely unknown, so the intelligent system estimates the complete model, such as

$$\hat{F}_2(x_e, \Theta) \approx F_0(x_e) + \Delta F(x_e)$$

Therefore u and ω in Equation (8) is rewritten as

$$u = \hat{F}_2(x_e, \Theta) - \tau$$

$$\omega = (F(x_e) - \hat{F}_2(x_e, \Theta)) + \tau_d$$

so that ω includes the estimation error from the intelligent system.

Regarding the nonlinear $\mathcal{H}_\infty$ control solution proposed by Chang and Chen [24] for constrained systems, define $u = \bar{u}$ where $\bar{u} = u_P + u_F$, such that

$$u_P = k_0 E \bar{x}_2 - k(x_e) sgn(L\bar{x}_2) \tag{10}$$

$$u_F = J_c^T \lambda_c \tag{11}$$

where u_P is the $\mathcal{H}_\infty$+VSC control term for the position enforcement and u_F is the H_∞ control law for the force tracking procedure, with

$$E := \begin{bmatrix} I_{(n-m)} \\ 0_{m\times(n-m)} \end{bmatrix} \quad \text{and} \quad \lambda_c \doteq \lambda_d - k_\lambda \int_0^T (\lambda - \lambda_d)dt$$

for some constant gain k_0, $k(x_e) > 0$ and $k_\lambda > 0$.

Theorem 1 presented in the following is a variation of the results presented in [21,23,24] with the difference that the nominal model and an intelligent system are considered rather than the linear parametrization of the robot. Thus, this theorem defines an adaptive H_∞ controller to solve the same problem's position/force tracking control problems presented in these papers but with the advantage of using the known model of the robot.

Theorem 1 *Given a desired disturbance attenuation level* $\gamma > 0$*, a weighting matrix Q and the Lyapunov candidate function V(t), the following performance criterion*

$$\int_0^T \|\bar{x}(t)\|_Q^2 \leq V(0) + \gamma^2 \int_0^T \|\omega(t)\|^2, \quad \forall T \geq 0 \tag{12}$$

is satisfied, for $\omega(t) \in L_2\,[0,\infty)$, if there exists a dynamic state feedback controller

$$\dot{\Theta} = \begin{cases} -\rho^{-T}\Xi^T L\bar{x}_2 & \text{if } \|\Theta\| < M_\theta \text{ or } (\|\Theta\| = M_\theta \text{ and } \bar{x}_2^T L^T \Xi\Theta \geq 0) \\ -\rho^{-T}\Xi^T L\bar{x}_2 + \rho^{-T}\frac{\bar{x}_2^T L^T \Xi\Theta}{\|\Theta\|^2}\Theta & \text{if } \|\Theta\| = M_\theta \text{ and } \bar{x}_2^T L^T \Xi\Theta < 0 \end{cases} \tag{13}$$

$$\tau = F_0(x_e) + \Xi\Theta - k_0 E\bar{x}_2 + k(x_e)sgn(L\bar{x}_2) - J^T\lambda_c \tag{14}$$

which is the solution of the adaptive nonlinear H_∞ control problem subject to Equation (6) for the cases 1 and 2.

The stability proof of this dynamic controller follows the line of the proof presented by Chang and Chen [24].

6. FORCE/MOMENT SENSOR DEVELOPMENT

The controllers proposed in Section 5 require the inferred force and moment measures in the robotic end-effector. The most accurate way to obtain such information is by measurements obtained through a multi-axes load cell.

This section presents a detailed description on the modular 3D dynamic force/moment sensor device developed for this application purpose. Its modular architecture is shown in Figure 3.

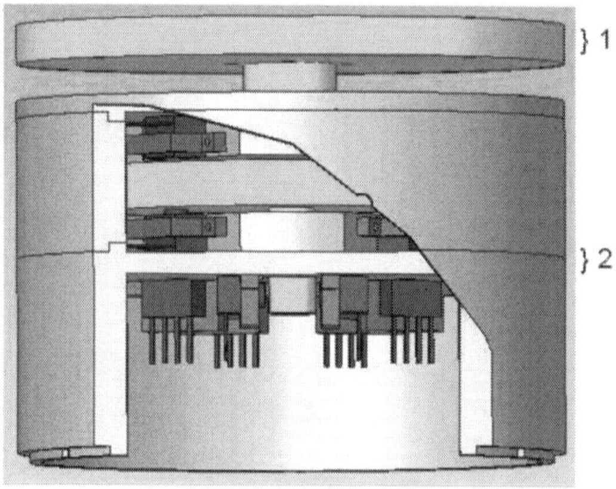

Figure 3. Design of the 3D dynamic forces/moments sensor: (1) Moving part; (2) Fixed part.

6.1. Mechanical Description

The 3D dynamic forces/moments sensor essentially comprised of two parts: a moving part and a fixed part. The moving part is shown in yellow in Figure 3(1). It is composed of a base (outside the sensor body) and a body of force transmission (inside the sensor body), which is responsible for transmitting forces and moments to the unidirectional force sensor units mounted on the fixed part, Figure 3(2). The device has an aluminum-made sensor board where twelve force sensor units are disposed in order to measure force and moment in all directions. The sensor units are arranged in pairs, for example, sensors 1A and 1B as shown in Figure 4, which illustrates the body of forces transmission (4) and identifies the sensor units.

Each pair of sensor units measures the force in one direction. This setting is tensioned proportionally to the movement performed by (4) in that axis. Therefore, forces and moments normal to the plane of the base are calculated by the composition of measured forces in the three axes.

Figure 5 demonstrates the sensor units operation schematic, and identifies the force measurements.

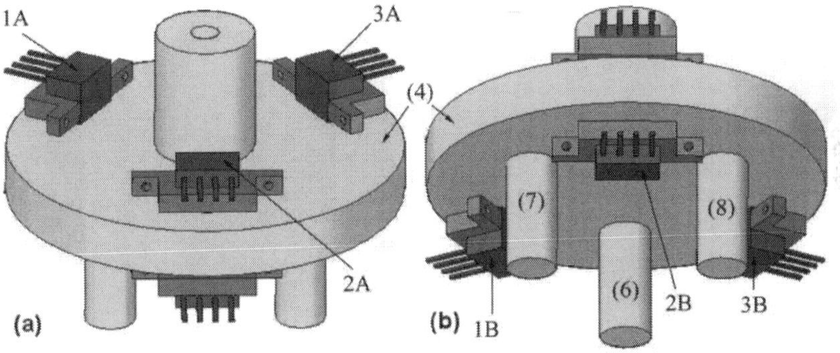

Figure 4. Body of force transmission (4) and arrangement of sensor units: (**a**) Top view; (**b**) Bottom view.

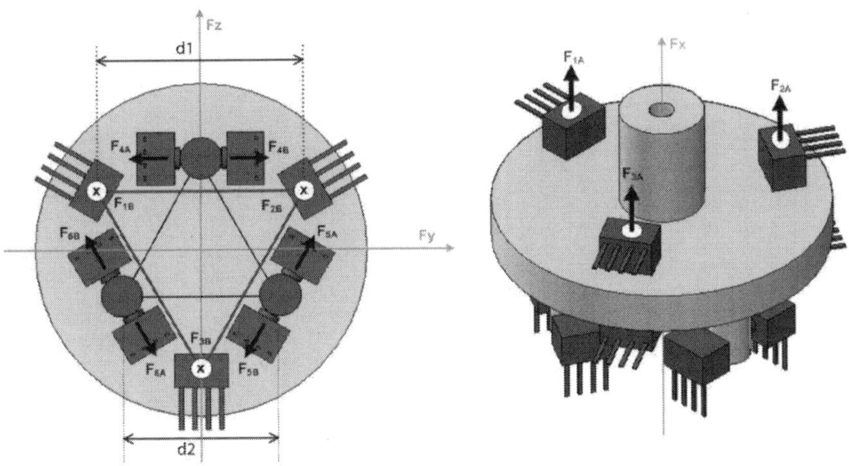

Figure 5. Schematic of sensor operation.

The formulation of the resultant forces/moments composition is obtained as follows.

Three pairs of sensor units, (1A,1B), (2A,2B) and (3A,3B), define, respectively, the force measurements (F_{1A},F_{1B}), (F_{2A},F_{2B}) and (F_{3A},F_{3B}). The resultant forces of these three points can be expressed as

$$F_1 = F_{1A} - F_{1B} \tag{15}$$

$$F_2 = F_{2A} - F_{2B} \tag{16}$$

$$F_3 = F_{3A} - F_{3B} \tag{17}$$

It can be seen in Figure 5 that F_1, F_2, and F_3 form an equilateral triangle, whose side is given by d_1. In [29], Doebelin demonstrated that if three load cells, measuring force in one direction, are arranged triangularly as the vertices of the triangle (the measurement direction normal to the plane formed by the triangle), the following physical quantities can be calculated:

$$F_x = F_1 + F_2 + F_3 \tag{18}$$

$$M_y = d_1 \frac{F_1 - F_2 - (F_3 - F_1)}{2\sqrt{3}} \tag{19}$$

$$M_z = \frac{d_1(F_3 - F_2)}{2} \tag{20}$$

where F_x, M_y, and M_z are the force along x-axis and the moments along axes y and z, respectively.

In the same way, F_4, F_5, and F_6 also form an equilateral triangle, whose side is given by d_2. Thus, the resultant forces of these three points can be expressed as

$$F_4 = F_{4A} - F_{4B} \tag{21}$$

$$F_5 = F_{5A} - F_{5B} \tag{22}$$

$$F_6 = F_{6A} - F_{6B} \tag{23}$$

and therefore,

$$F_y = \frac{F_5 - F_4 - (F_4 - F_6)}{2} \tag{24}$$

$$F_z = \frac{\sqrt{3}(F_5 - F_6)}{2} \tag{25}$$

$$M_x = \frac{-d_2(F_4 + F_5 + F_6)}{2\sqrt{3}} \tag{26}$$

where F_y, F_z, and M_x are the forces along axes y and z and the moment along axis x, respectively. The sensor parts and the assembled device is shown in Figure 6.

The applied sensor unit provides a measurement range of 0 to 1.5 kgf, or a maximum load of 14.7 N. By means of Equations (18)–(20) and (24)–(26), the built sensor presents measurements characteristics as shown in Table 1.

Table 1. Maximum load.

	Axis X	Axis Y	Axis Z
Force (N)	44.1	29.4	25.5
Moment (N/m)	$d_2 12.7$	$d_1 17$	$d_1 14.7$

Figure 6. 3D dynamic forces/moments sensor.

6.2. Electronic Description

The signal from each pair of unidirectional sensors are conditioned by an electronic circuit based on Instrumentation Amplifiers (INA). The electrical schematic of the signal conditioning circuit is summarized in Figure 7.

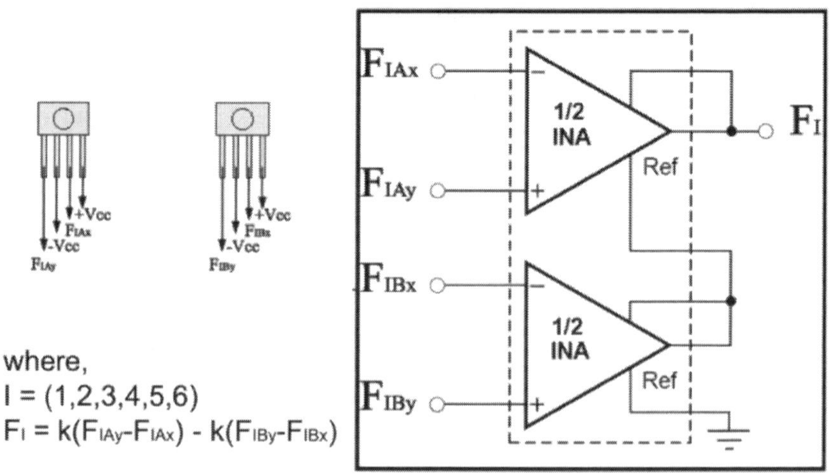

Figure 7. Electronic diagram.

The sensors are powered by a symmetric supply (+ Vcc and −Vcc) and provide differential output signals (for example, F_{1Ax} and F_{1Ay}).

Six INA comprise the printed circuit board (PCB). Each INA is responsible for conditioning output signals from a pair of force sensors. With this configuration it became possible to measure the differential forces applied on the moving part of the proposed device. The resultant force is then given by the following equation as described in Figure 7.

$$F_I = kF_{IA} - kF_{IB}$$

The signals generated by the INAs are directed to a data acquisition board that interfaces with the computer.

7. EXPERIMENTAL RESULTS

The designed 3D dynamic forces/moments sensor operation is validated in a set of experimental applications, where a three-link planar manipulator is subject to a holonomic constraint defined by a straight rule. The proposed intelligent adaptive robust controllers are applied to this plant for position and force trajectory tracking. The experimental manipulator UARM (UnderActuated Robot Manipulator), whose nominal parameters are given in Table 2, is composed of a DC motor in each joint, a break and an optical encoder with quadrature decoding used to measure joint positions, Figure 8(a). Joint velocities are obtained by numerical differentiation and filtering [30,31]. Modeling matrices for the UARM, M(q), C(q, $\dot{q}$), and G(q), can be seen in [30]. In Figure 8(b), the designed force sensor is coupled to the UARM end-effector which is constrained in a straight rule. A graphical user interface, whose software was developed using MATLAB® platform, is used to implement control laws and interface with the experimental setup. It was implemented in a modular form, since any new controller can be easily developed and applied.

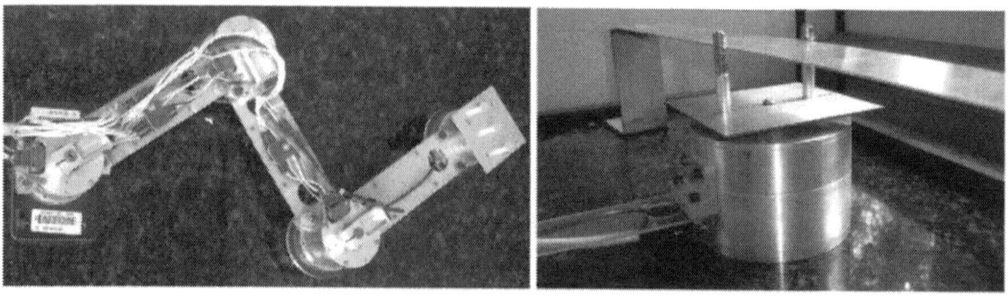

(a) (b)

Figure 8. (**a**) UnderActuated Robot Manipulator; (**b**) Force sensor device coupled to the UARM end-effector.

Table 2. UARM Parameters.

Body	m_i (kg)	I_i (kgm^2)	l_i (m)	lc_i (m)
Link 1	0.850	0.0075	0.203	0.096
Link 2	0.850	0.0075	0.203	0.096
Link 3	1.700	0.0900	0.240	0.177

7.1. Implementation of Control Law

The constraint surface for the robot end-effector is a segment of a straight line on the X–Y plane. The angle α between the end-effector and the constraint line is defined in this application example as π/2. It means that the orientation must remain in a constant value c given through the line inclination β and α, where $c = \pi + atan(\beta) - \alpha$. Hence, the equation of the m = 2 constraints is given by

$$\phi(q) = \begin{bmatrix} -l_1 s_1 - l_2 s_{12} - l_3 s_{123} + \beta[l_1 c_1 + l_2 c_{12} + l_3 c_{123}] + b \\ q_1 + q_2 + q_3 - c \end{bmatrix} = \begin{bmatrix} 0 \\ 0 \end{bmatrix}$$

where b is the linear coefficient of the constraint line, $s_{ij} = sin(q_i + q_j)$, $c_{ij} = cos(q_i + q_j)$. Hence, $\phi : \Re^3 \rightarrow \Re^2$, and the Jacobian matrix, $J_c(q) = \partial\phi/\partial q$, is given by

$$J_c = \begin{bmatrix} J_{c_{11}} & J_{c_{12}} & J_{c_{13}} \\ J_{c_{21}} & J_{c_{22}} & J_{c_{23}} \end{bmatrix}$$

with

$$\begin{aligned} J_{c_{11}} &= l_1 c_1 + l_2 c_{12} + l_3 c_{123} + \beta[l_1 s_1 + l_2 s_{12} + l_3 s_{123}] \\ J_{c_{12}} &= l_2 c_{12} + l_3 c_{123} + \beta[l_2 s_{12} + l_3 s_{123}] \\ J_{c_{13}} &= l_3 c_{123} + \beta[l_3 s_{123}] \\ J_{c_{21}} &= J_{c_{22}} = J_{c_{23}} = 1 \end{aligned}$$

Defining $q^1 = [q_1]$ and $q^2 = [q_2 \;\; q_3]$, the matrix L(q) of the constraint line is

$$L(q) = \begin{bmatrix} 1 \\ -\frac{[l_1 \cos(q_1) + l_2 \cos(q_1+q_2) + \beta(l_1 \sin(q_1) + l_2 \sin(q_1+q_2))]}{l_2[\cos(q_1+q_2) + \beta \sin(q_1+q_2)]} \\ \frac{[l_1 \cos(q_1) + l_2 \cos(q_1+q_2) + \beta(l_1 \sin(q_1) + l_2 \sin(q_1+q_2))]}{l_2[\cos(q_1+q_2) + \beta \sin(q_1+q_2)]} - 1 \end{bmatrix}$$

Initial and final coordinates of the movement are $(x_0, y_0) = (0.46, 0.38)$ m and $(x(T), y(T)) = (0.53, 0.13)$ m, respectively. In this case, $\beta = -3.57, b = 2.02$, and $c = 15.6°$. The reference trajectory for the joint variables $q_d(t) = q_d^1(t)$ is a fifth-degree polynomial, with trajectory duration time T = 4 s. It is desired that no force acts on the normal direction of the constrain line and no moment acts on the z-direction, that is, $\lambda_d = [(F_x)_d \;\; (M_z)_d]^T = [0 \;\; 0]^T$.

We use as benchmarking for comparison of the proposed controllers, a controller with no intelligent adjust adaptation, that is, only the nominal parameters are used. In other words, we use the controller of Equation (14) without the ΞΘ term.

During the experiment, a limited disturbance was introduced at $t_i = 1.0$s in the following form:

$$\tau_d = \begin{bmatrix} 0,01e^{\frac{-(t-t_d)^2}{2\mu^2}} \text{sen}(3,6\pi t) \\ -0,01e^{\frac{-(t-t_d)^2}{2\mu^2}} \text{sen}(2,7\pi t) \\ 0,01e^{\frac{-(t-t_d)^2}{2\mu^2}} \text{sen}(1,8\pi t) \end{bmatrix}$$

If compared with the nominal torque, the disturbance τ_d is approximately 64% of its peak value, see Figure 9.

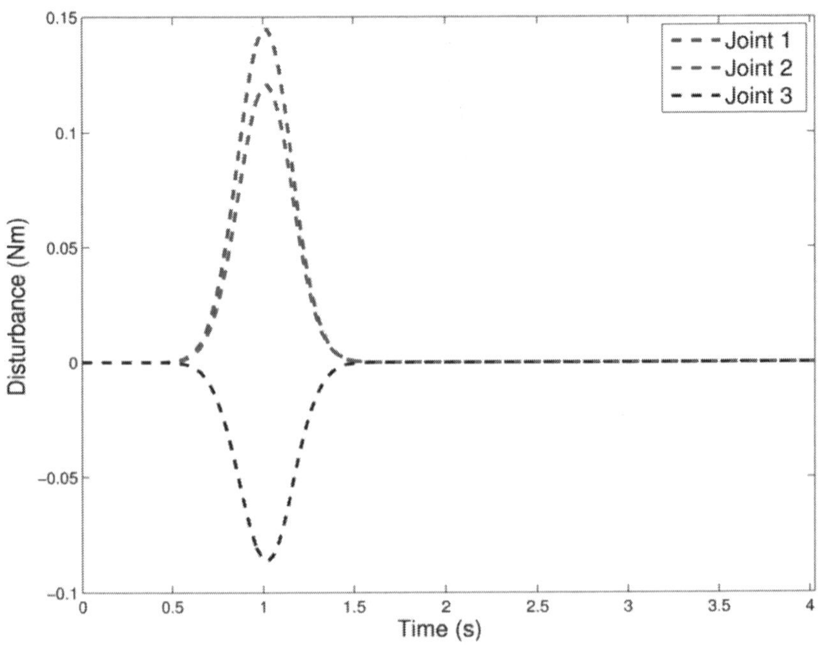

Figure 9. Torque disturbance.

The selected gains are defined in Table 3.

Table 3. Selected Gains.

P	k_O	k_λ	ρ
$\begin{bmatrix} 2.50 & 0 & 0 \\ 0 & 2.50 & 0 \\ 0 & 0 & 2.50 \end{bmatrix}$	$\begin{bmatrix} 0.35 & 0 & 0 \\ 0 & 0.35 & 0 \\ 0 & 0 & 0.35 \end{bmatrix}$	$\begin{bmatrix} 2.00 & 0 & 0 \\ 0 & 2.00 & 0 \\ 0 & 0 & 2.00 \end{bmatrix}$	$\begin{bmatrix} 0.75 & 0 & 0 \\ 0 & 0.75 & 0 \\ 0 & 0 & 0.75 \end{bmatrix}$

7.1.1. Neural Networks Configuration

For the nonlinear H_∞ controllers via neural network, Cases 1 and 2, proposed in Section 5, let n = 3 be the number of joints of manipulator. Define

$$\hat{F}_{NNi}(x_e,\Theta) = [\ \hat{F}_{NNi_1}(x_e,\Theta_1)\ \ \hat{F}_{NNi_2}(x_e,\Theta_2)\ \ \hat{F}_{NNi_3}(x_e,\Theta_3)\ ]^T = \Xi\Theta$$

with $p_1 = p_2 = p_3 = 7$ neurons in the hidden layer and the bias vector $b_1 = b_2 = b_3 = [\ -3\ \ -2\ \ -1\ \ 0\ \ 1\ \ 2\ \ 3\ ]$. Using the reduced model, the input vector x_e is defined by *xe*= $[\ q^1\ \ \dot{q}^1\ \ q_d^1\ \ \dot{q}_d^1\ \ \ddot{q}_d^1\ ]$. In order to provide an input related to the error values, the weighting matrix for the first layer of the neural networks is defined $W_i^1 = W_i^2 = W_i^3 = [w_{ij}^1] = [w_{ij}^2] = [w_{ij}^3]\ \ [\ 1\ \ 1\ \ -1\ \ -1\ \ -1\ ]$. The activation function of the neurons in the hidden layer is chosen as the hyperbolic tangent, $G(.) = tanh(.)$. The uncertain vector Θ is defined as $\Theta = [\ \Theta_1^T\ \ \Theta_2^T\ \ \Theta_3^T\ ]^T$ and the matrix $\Xi = [\ \xi_1^T\ \ \xi_2^T\ \ \xi_3^T\ ]^T$ as Pazelli et al. and Siqueira et al. showed in [28,30]. Θ values are updated at each control iteration by the adaptive law given in Equation (13). The neural networks outputs are given by

$$\hat{F}_{NNi_k}(x_e,\Theta_k) = \sum_{i=1}^{p_k} \theta_{ki} G\left(\sum_{j=1}^{q_k} w_{ij}^k x_{ej} + b_i^k\right) = \xi_k^T \Theta_k$$

Experimental results for the Case 1 (NN1) with neural network are shown in Figures 10(a), 11(a), and 12(a). For the Case 2 with neural network plus nominal model (NN2), experimental results are shown in Figures 10(b), 11(b), and 12(b).

Remark 2 Although NN1 and NN2 use the same structure and the same input data, their outputs are different. The online adaptation law provides the update of Θ values in distinct ways for each case. The contribution of NN1 output to the final torque signal is lower than the contribution of NN2 output, since NN1 estimates the value of $\Delta F(x_e)$ and NN2 estimates the value of $F(x_e) + \Delta F(x_e)$. Thus, Θ values must achieve that requirement for each case.

7.1.2. Fuzzy Systems Configuration

For the proposed adaptive fuzzy nonlinear H_∞ controllers, Cases 1 and 2 (Section 5), a set of n = 3 fuzzy systems may be defined by

$$\hat{F}_{FSi}(x_e, \Theta) = [\ \hat{F}_{FSi_1}(x_e, \Theta_1) \quad \hat{F}_{FSi_2}(x_e, \Theta_2) \quad \hat{F}_{FSi_3}(x_e, \Theta_3)$$

where $\hat{F}_{FSi_1}(.,.)$, $\hat{F}_{FSi_2}(.,.)$, and $\hat{F}_{FSi_3}(.,.)$ correspond to the estimate of the uncertain part of the dynamic behavior of joints 1, 2, and 3, respectively. The input vector x_e is defined as $x_e = \begin{bmatrix} \tilde{q}_1 & \dot{\tilde{q}}_1 \end{bmatrix}$. The fuzzy sets $A(x_e) = \begin{bmatrix} A_1(\tilde{q}_1) & A_2(\dot{\tilde{q}}_1) \end{bmatrix}$ are defined according to Figure 13 and the number of linguistic variables in U_1 and U_2 are defined as $r_1 = r_2 = 3$. They are applied to both $\hat{F}_{FSi_1}(.,.)$, $\hat{F}_{FSi_2}(.,.)$ and $\hat{F}_{FSi_3}(.,.)$ for the universe of discourse of position errors, $u_1^1 = u_1^2 = u_1^3 = \tilde{q}_1 \in U_1^1 = U_1^2 = U_1^3 = U_1$, and for the universe of discourse of velocity errors, $u_2^1 = u_2^2 = u_2^3 = \dot{\tilde{q}}_1 \in U_2^1 = U_2^2 = U_2^3 = U_2$.

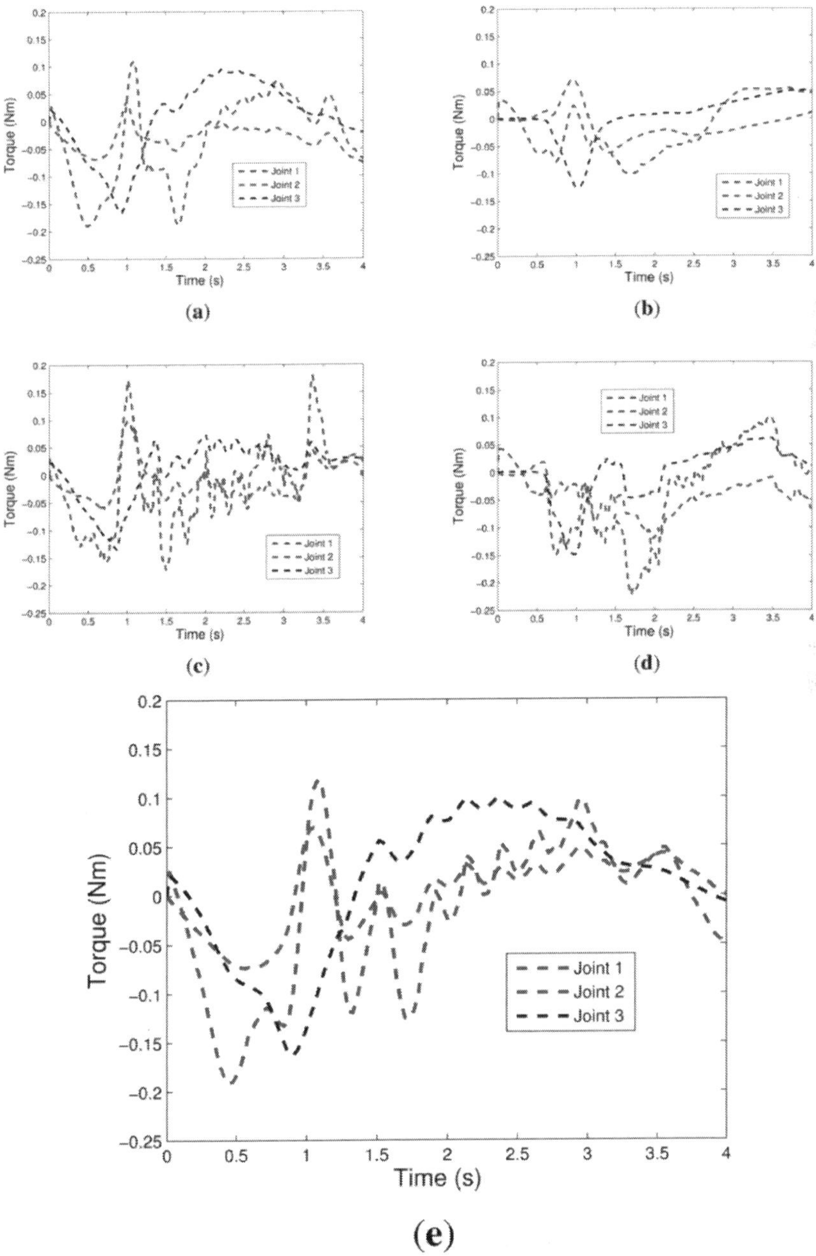

Figure 10. Joint Torque. (**a**) NN1; (**b**) NN2; (**c**) FS1; (**d**) FS2; (**e**) NOM.

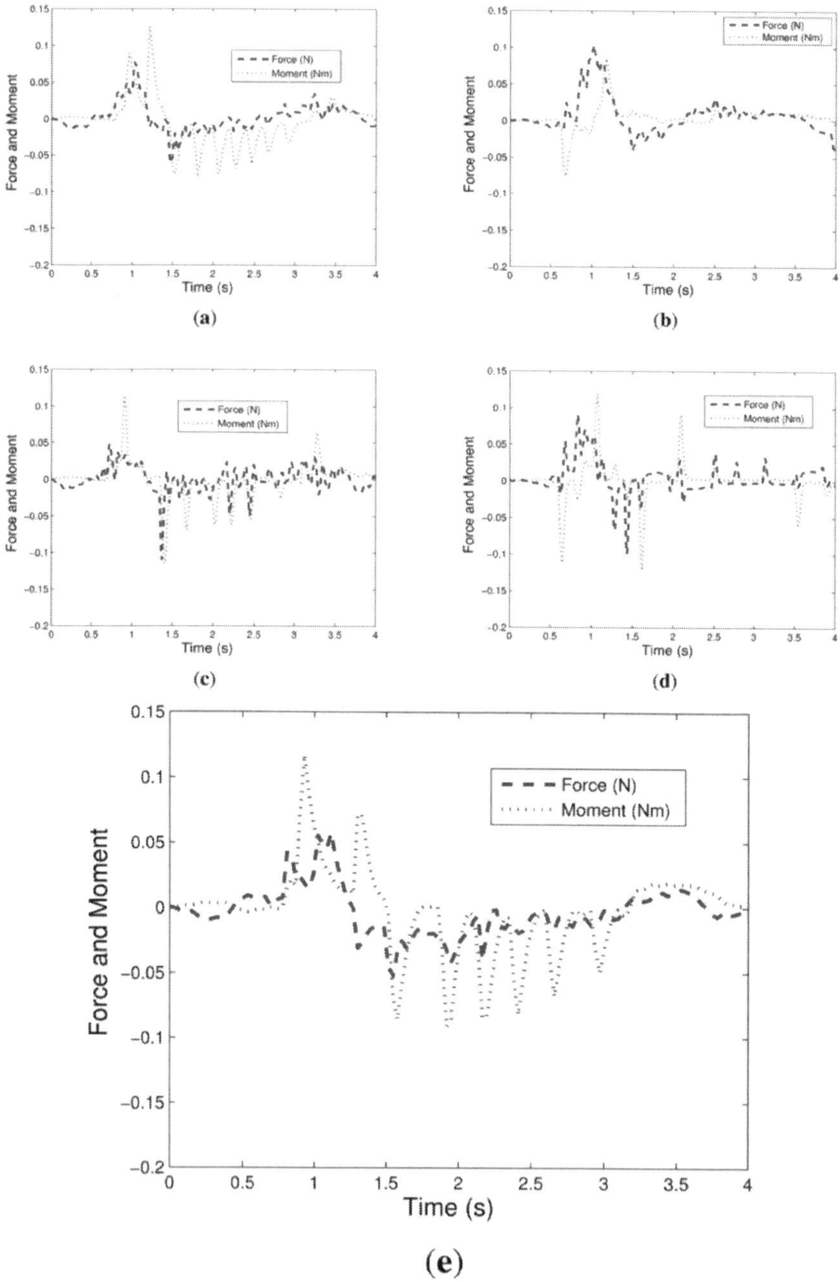

Figure 11. Force and Moment. (**a**) NN1; (**b**) NN2; (**c**) FS1; (**d**) FS2; (**e**) NOM.

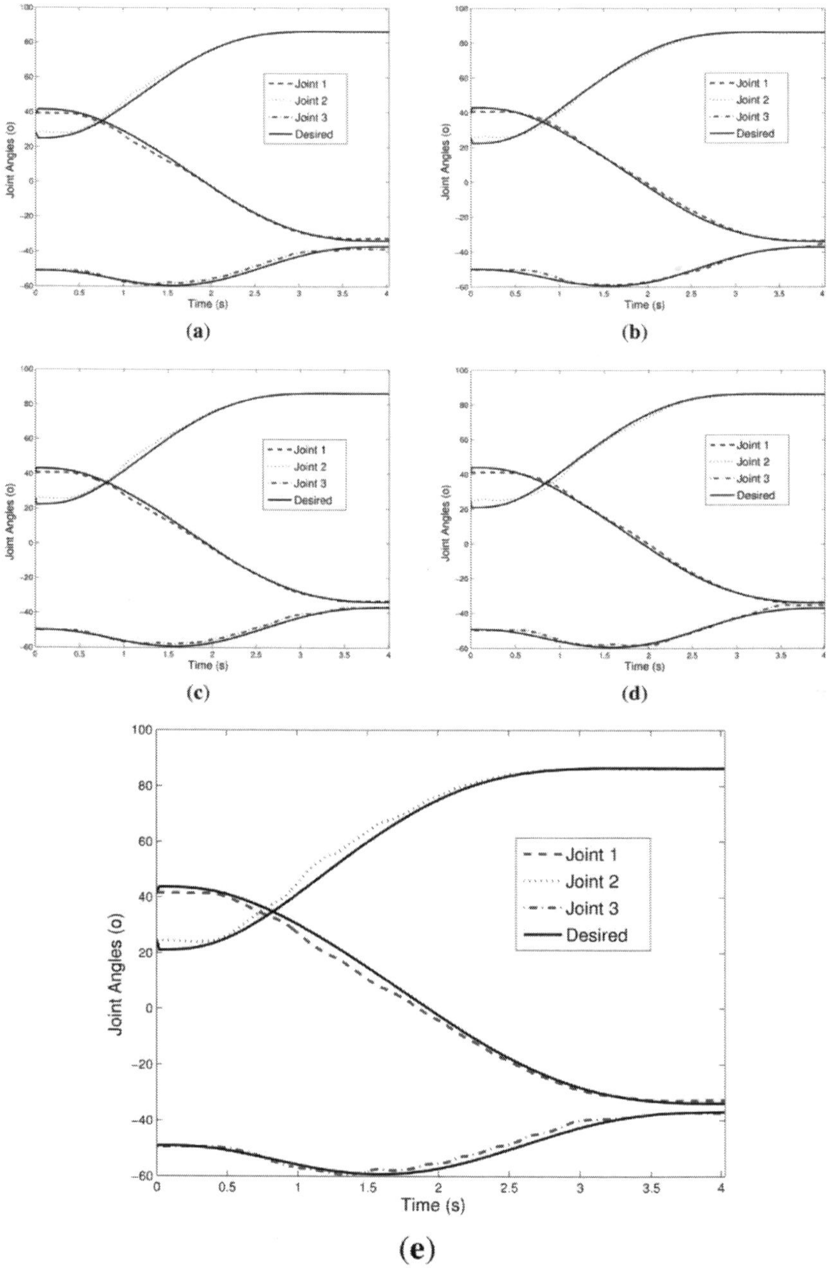

Figure 12. Joint Angles. (**a**) NN1; (**b**) NN2; (**c**) FS1; (**d**) FS2; (**e**) NOM.

During the control loop, these graphics are used to determine the grade of membership associated to x_e, which defines the value of μ in the output of the Takagi–Sugeno fuzzy model; for more details see [28],

$$\hat{F}_{FSi_k}(x_e,\Theta_k) = \frac{\sum_{i=1}^{p_k}\mu_i^k y_i^k}{\sum_{i=1}^{p_k}\mu_i^k} = \frac{\sum_{i=1}^{p_k}\mu_i^k(\theta_{i0}^k + \theta_{i1}^k u_1 + \theta_{i2}^k u_2)}{\sum_{i=1}^{p_k}\mu_i^k} = \xi_k^T\Theta_k$$

A fuzzy rule base is defined with $p_k = r_1 r_2 = 9$ rules as

R_i : **IF**$(u_1$ is $A_{r_{1i}1})$ and $(u_2$ is $A_{r_{2i}2})$ **THEN** y_i

The vector of adjustable components Θ is defined as $\Theta = [\ \Theta_1^T\ \ \Theta_2^T\ \ \Theta_3^T\ ]^T$, with

$$\begin{aligned}
\Theta_1^T &= [\ \theta_{10}^1\ \ \theta_{11}^1\ \ \theta_{12}^1\ \ \cdots\ \ \theta_{90}^1\ \ \theta_{91}^1\ \ \theta_{92}^1\ ]\\
\Theta_2^T &= [\ \theta_{10}^2\ \ \theta_{11}^2\ \ \theta_{12}^2\ \ \cdots\ \ \theta_{90}^2\ \ \theta_{91}^2\ \ \theta_{92}^2\ ]\\
\Theta_3^T &= [\ \theta_{10}^3\ \ \theta_{11}^3\ \ \theta_{12}^3\ \ \cdots\ \ \theta_{90}^3\ \ \theta_{91}^3\ \ \theta_{92}^3\ ]
\end{aligned}$$

and are updated at each control iteration by the adaptive law in Equation (13).

The matrix $\Xi = [\ \xi_1^T\ \ \xi_2^T\ \ \xi_3^T\ ]^T$ is computed with

$$\begin{aligned}
\xi_1^T &= [\ \xi_{10}^1\ \ \xi_{11}^1\ \ \xi_{12}^1\ \ \cdots\ \ \xi_{90}^1\ \ \xi_{91}^1\ \ \xi_{92}^1\ ]\\
\xi_2^T &= [\ \xi_{10}^2\ \ \xi_{11}^2\ \ \xi_{12}^2\ \ \cdots\ \ \xi_{90}^2\ \ \xi_{91}^2\ \ \xi_{92}^2\ ]\\
\xi_3^T &= [\ \xi_{10}^3\ \ \xi_{11}^3\ \ \xi_{12}^3\ \ \cdots\ \ \xi_{90}^3\ \ \xi_{91}^3\ \ \xi_{92}^3\ ]
\end{aligned}$$

Figures 10(c), 11(c), and 12(c) show the experimental results for Case 3. Figures 10(d), 11(d), and 12(d) show the experimental results of fuzzy system plus nominal model for Case 4.

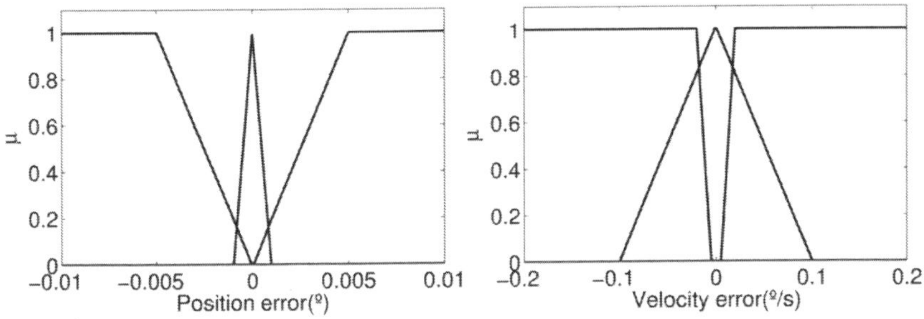

Figure 13. Fuzzy sets $A_1(\tilde{q}_1)$ and $A_2(\tilde{q}_1)$.

7.2. Results Discussion

Figure 10 shows the applied torques in the robotic manipulator joints. Note that for the range of the disturbance inserted, the controllers act strongly on the system, reversing the directions of the torques of joints 1 and 2 and increasing the torque of joint 3. In this analysis, it appears that the controllers using neural networks, NN1 and NN2, oscillate less, therefore, they have a slightly slower response time.

The charts of Figure 11 are important because they show the behavior of the end effector during the controller actions, and represent the measurements of force and moment in the robotic end-effector. As expected, it was observed that during the period in which the disturbance appeared in the system, there was a higher intensity of forces and moments which decreased gradually, tending to zero, until the final time (4 s). The proposed configuration for the device sensor proves its effectiveness in providing these data with the required accuracy and without delay to the implementation of the control laws under study.

Figure 12 shows the tracking angles of three joints of the robotic manipulator. Notice that the largest deviation is in the range where the disturbance appeared. Nonetheless, the experiment showed good tracking of the trajectories desired. Again, we can see that for the controller without the intelligent system, NOM, the control actuation is slower, providing greater tracking error of reference signals.

Three performance indexes are used to compare the nonlinear H_∞ controllers: the L_2 norm of the state vector

$$\mathcal{L}_2[\tilde{x}] = \left(\frac{1}{(t_r - t_0)} \int_{t_0}^{t_r} \|\tilde{x}(t)\|_2^2 \, dt \right)^{1/2}$$

where $\| \cdot \|_2$ is the Euclidean norm, the sum of the applied torques

$$E[\tau] = \sum_{i=1}^{3} \left(\int_{t_0}^{t_r} |\tau_i(t)| \, dt \right)$$

and the sum of areas of contact forces

$$E[\lambda] = \sum_{i=1}^{3} \left(\int_{t_0}^{t_r} |\lambda_i(t)| \, dt \right)$$

where $\lambda_i(t)$ the ith component of the contact forces. As the desired values of contact forces are zero, the lower the value of E $[\lambda]$, the better the controller will be with respect to contact force control.

The results that are shown in Table 4 present the mean value of five experiments.

Table 4. Performance Indexes.

	$\mathcal{L}_2[x]$ (m)	$E[\tau]$ (Nms)	$E[\lambda]$ (Ns)
Nominal	0.1031	0.6508	0.1506
NN1	0.0780	0.6564	0.1435
NN2	0.1026	0.4318	0.1156
FS1	0.0759	0.5865	0.1136
FS2	0.0899	0.5182	0.1061

From Table 4 we can conclude that the controller based only on the nominal model presents a higher state error and greater forces. Although the fuzzy controllers exhibit the best performance, this performance difference with respect to controllers based on neural networks deserves more in-depth study, which will be carried out in future works.

8. CONCLUSIONS

In this work, the control problem of trajectory tracking with H_∞ guaranteed performance was considered for manipulators with force and position constraint. Five controllers were evaluated:

- The first controller, called nominal, considers that the term $F(x_e)$ is completely known, i.e., it does not take into account parametric uncertainties of the model.
- The second and third controllers have neural network-based estimators, where the second considers the known nominal model of the manipulator and estimates only parameter uncertainties, and the third estimates the full model.
- The fourth and fifth controllers have fuzzy logic-based estimators which estimate only parametric uncertainties and the full model, respectively.

According to experimental results presented in Section 7, controllers NN1 and FS1 have better tracking trajectory compared, respectively, with the NN2 and FS2. Moreover, it also appears they are more stable because of the lower oscillations after the introduction of disturbance, and this feature occurs for the trajectory tracking in Cartesian space in order to follow the trajectory of joint angles. Comparing the results of intelligent systems it is noted that the controllers via neural networks (NN2 and NN1) have a smoother control action and, consequently, have smaller oscillations than the fuzzy systems (FS1 and FS2), which can be verified by comparing the graphics of torque (Figure 10). On the other hand, the controllers NN1 and NN2 are slower than the FS1 and FS2, hence they tend to have a greater tracking error. In addition, when the system is subject to disturbances, all controllers respond well.

Analyzing the proposed performance indices, it was shown that the state error in the fuzzy system-based controllers (FS1 and FS2) tend to be smaller than those based on neural networks (NN2 and NN1). A possible explanation for this is due to the fact that fuzzy systems acted faster than neural networks. It was observed that choosing values of k_λ more directly influence the adjustment of the forces, while the values of ρ influence the tracking errors in the state variables.

The development of the proposed sensor was of crucial importance for the realization of the experimental analysis, a differential of this paper. The measurement results shown proved the effectiveness of the proposed structure. Still, its modular feature enables the application of sensor units of various kinds, such as piezoelectric sensors.

As future work, a comparative study will be performed between the developed sensor and commercially available ones. Furthermore, the use of genetic algorithms to adjust controller gains is a current and ongoing project.

ACKNOWLEDGMENTS

This work was supported by Coordenação de Aperfeiçoamento de Pessoal de Nível Superior (CAPES) and Fundação de Apoio à Pesquisa do Estado de São Paulo (FAPESP), under grant 2012/05552-9.

REFERENCES

1 Kilicaslan, S.; Özgören, M.K.; Ider, K. Hybrid force and motion control of robots with flexible links. Mech. Mach. Theory **2010**, 45, 91–105.

2 Lewis, F.L.; Abdallah, C.T.; Dawson, D.M. Control of Robot Manipulators; MacMillan Publishing Company: New York, NY, USA, 1993.

3 Bechlioulis, C.P.; Doulgeri, Z.; Rovithakis, G.A. Neuro-adaptive force/position control with prescribed performance and guaranteed

contact maintenance. IEEE Trans. Neural Netw. **2010**, 21, 1857–1868.

4 Cheah, C.C.; Hou, S.P.; Zhao, Y.; Slotine, J.J.E. Adaptive vision and force tracking control for robots with constraint uncertainty. IEEE/ASME Trans. Mechatron. **2010**, 15, 389–399.

5 Karayiannidis, Y.; Doulgeri, Z. Adaptive control of robot contact tasks with on-line learning of planar surfaces. Automatica **2009**, 45, 2374–2382.

6 Karayiannidis, Y.; Doulgeri, Z. Robot contact tasks in the presence of control target distortions. Robot. Auton. Syst. **2010**, 58, 596–606.

7 Kumar, N.; Panwar, V.; Sukavanam, N.; Borm, S.P.S.J.H. Neural network based hybrid force/position control for robot manipulators. Int. J. Precis. Eng. Manuf. **2011**, 12, 419–426.

8 Salisbury, J.K. Active Stiffness Control of a Manipulator in Cartesian Coordinates. Proceedings of the 19th IEEE Conference on Decision and Control including the Symposium on Adaptive Processes, Albuquerque, NM, USA, 10–12 December 1980; Volume 19, pp. 95–100.

9 Raibert, M.; Craig, J. Hybrid position/force control of manipulators. ASME J. Dyn. Syst. Meas. Control **1981**, 102, 126–133.

10 Paul, R.; Shimano, B. Compliance and Control. Proceedings of the Joint Automatic Control Conference, Purdue, IN, USA, 27–30 July 1976; pp. 694–699.

11 McClamroch, N.H.; Wang, D. Feedback stabilization and tracking of constrained robots. IEEE Trans. Autom. Control **1988**, 33, 419–426.

12 Ortega, R.; Spong, M.W. Adaptive motion control of rigid robots: A tutorial. Automatica **1989**, 25, 877–888.

13 Slotine, J.J.E.; Li, W. Composite adaptive control of robot manipulators. Automatica **1989**, 25, 509–519.

14 Slotine, J.J.E.; Li, W. Applied Nonlinear Control; Prentice-Hall: Englewoods Cliffs, NJ, USA, 1991.

15 Spong, M.W.; Vidyasagar, M. Robot Dynamics and Control; John Wiley and Sons: New York, NY, USA, 1989.

16 Carelli, R.; Kelly, R. An adaptive impedance/force controller for robot manipulators. IEEE Trans. Autom. Control **1991**, 36, 967–971.

17 Jean, J.H.; Fu, L.C. Adaptive hybrid control strategies for constrained robots. IEEE Trans. Autom. Control **1993**, 38, 598–603.

18 Panteley, E.V.; Stotsky, A.A. Adaptive trajectory/force control scheme for constrained robot manipulators. Int. J. Adapt. Control Signal Process. **1993**, 7, 489–496.

19 Su, C.Y.; Stepanenko, Y.; Leung, T.P. Combined adaptive and variable structure control for constrained robots. Automatica **1995**, 31, 483–488.

20 Su, C.Y.; Leung, T.P.; Zhou, Q.J. Force/motion control of constrained robots using sliding mode. IEEE Trans. Autom. Control **1992**, 37, 668–672.

21 Chang, Y.C.; Chen, B.S. Adaptive tracking control design of constrained robot systems. Int. J. Adapt. Control Signal Process. **1998**, 12, 495–526.

22 Chang, Y.C.; Chen, B.S. A nonlinear adaptive H_∞ tracking control design in robotic systems via neural networks. IEEE Trans. Control Syst. Technol. **1997**, 5, 13–28.

23 Chang, Y.C. Neural network-based H_∞ tracking control for robotic systems. IEEE Proc. Control Theory Appl. **2000**, 147, 303–311.

24 Chang, Y.C.; Chen, B.S. Robust tracking designs for both holonomic and nonholonomic constrained mechanical systems: Adaptive fuzzy approach. IEEE Trans. Fuzzy Syst. **2000**, 8, 46–66.

25 Meyer, R.A.; Lowe, A.E. Multiple Axis Load Sensitive Transducer. U.S. Patent 4640138, 1987.

26 Meyer, R.A.; Olson, D.J. Load Transducer. U.S. Patent 4821582, 1989.

27 Sommerfeld, J.L.; Meyer, R.A.; Larson, B.A.; Olson, D.J. Multi-Axis Load Cell. U.S. Patent 5969268, 1999.

28 Pazelli, T.F.P.A.T.; Terra, M.H.; Siqueira, A.A.G. Experimental investigation on adaptive robust controller designs applied to a free-floating space manipulator. Control Eng. Pract. **2011**, 9, 395–408.

29 Doebelin, E.O. Measurement Systems Application and Design; McGraw-Hill Science: New York, NY, USA, 1989; pp. 410–413.

30 Siqueira, A.A.G.; Terra, M.H.; Bergerman, M. Robust Control of Robots: Fault Tolerant Approaches; Springer: London, UK, 2011.

31 Siqueira, A.A.G.; Terra, M.H. Nonlinear and Markovian H_∞ controls of underactuated manipulators. IEEE Trans. Control Syst. Technol. **2004**, 12, 811–826.

CHAPTER 6

Robust Adaptive Neural Network Control for XY Table

Nguyen Hoang Giap[1], Jin-Ho Shin[2], Won-Ho Kim[2]

[1]Department of Intelligent System Engineering, Graduate School of Dong-eui University, Busan, South Korea

[2]Department of Mechatronics Engineering, Dong-eui University, Busan, South Korea

ABSTRACT

This paper proposed a robust adaptive neural network control for an XY table. The XY table composes of two AC servo drives controlled independently. The neural network with radial basis function is employed for velocity and position tracking control of AC servo drives to improve the system's dynamic performance and precision. A robust adaptive term is applied to overcome the external disturbances. The stability and the convergence of the system are proved by Lyapunov theory. The proposed controller is implemented in a DSP-based motion board. The validity and robustness of the controller are verified through experimental results.

KEYWORDS

Robust Adaptive Neural Network; Motion Control; XY Table; DSP

1. INTRODUCTION

The XY tables have taken an important role in manufacturing systems. Generally, the XY table is a ball-screw driven mechanism actuated by two AC servo drives. Such mechanism often exist many kinds of disturbances, nonlinear friction and uncertainties that limit the tracking performance of controlled system. In order to improve the performance of XY table, many works have been presented [1-3]. In [4], a combination of friction-modelbased feed-forward and inverse-model-based disturbance observe is used to reduce the radial tracking error and quadrant glitches of linear-drive XY table. This technique increases the system bandwidth, hence improves the overall tracking performance. However, robustness to variations of system parameters and uncertainties among each axis mechanism were not considered in the design of the compensation.

In most manufacturing processes, XY table often utilizes two AC servo motors and couples their output shafts to mechanical translators such as gears or bears to perform linear motions. Such mechanical systems have been extensively used in the industrial market due to low cost, high torque density, little torque ripple and power saving. Along with their intensive applications, many works have been conducted to enhance the performance of the motion control system [5-7]. In [8], the authors developed an auto-disturbance rejection controller based on a nonlinear tracking-differentiator, an extended states observer and a nonlinear PD control to achieve high performance and large robustness motion control. Although this simple method can archive high precision of AC servo system under unknown parameters and large load disturbance, the stability of the whole system is not guaranteed. In [9], a back propagation neural network is used to identify the dynamic parameters of AC servo motor. Although the controller obtained excellent performance compared with traditional PID controller, the load torque disturbance and uncertainties were not considered in the design.

In this paper, we propose a robust adaptive neural network (RANN) scheme to overcome such described drawbacks. The main aim of our controller is to define an augment error related to the angular position error and angular velocity error. Then the controller is designed so that the augment error can converge toward zero asymptotically. This process makes the angular velocity error and angular position error stable asymptotically. The controller is able to guarantee the convergence and stability of the servo system despite the existence of model uncertainties and external disturbances. In this controller, a 2-layer neural network with radial basis function is used to approximate the nonlinear factors of the AC servo motor. A robust adaptive control scheme is applied to suppress the disturbances and guarantee the stability of the system. The proposed robust adaptive neural network controller is implemented in a motion controller based on the high performance digital signal processor (DSP) TMS320C6727. The high speed DSP allows the closedloop controller and reference trajectory generators to be implemented in the DSP, while the feedback position signals with digital filter and digital inputs/outputs are carried out in the FPGA to reduce the computational load of DSP. Afterward, the output current control is put forward to the analog amplifier to drive the motor. This hardware structure allows for designing a highly sampled, flexible and compact motion controller.

This paper is organized as follows. Section 2 describes the development of robust adaptive neural network controller with the proof of asymptotic stability. Section 3 presents experimental results, including the comparison of the traditional PID controller and the proposed controller. Section 4 draws the conclusions.

2. DEVELOPMENT OF ROBUST ADAPTIVE NEURAL NETWORK CONTROLLER

2.1. Model of Two-Axis XY Table

The XY table used in this paper is composed of two AC servo drives controlled separately. Therefore, dynamic equation of a single-axis AC servo drive can be described in a rotating reference frame as following:

$$J\dot{\omega} + B\omega + \tau_d = \tau \tag{1}$$

where ω is angular velocity of the rotor, J is the moment of inertia, B is the viscous coefficient. τ is the driving torque, τ_d can be considered as the unknown load torque disturbance on the rotating shaft.

In the system design, it is assumed that the reference angular position is a continuous and differentiable signal. Therefore, the angular velocity and angular acceleration can be expressed as:

$$\omega = \dot{\theta} \tag{2}$$

$$\dot{\omega} = \ddot{\theta} \tag{3}$$

where θ is the angular position.

The goal of our control scheme is to achieve asymptotic tracking for a given continuous reference trajectory.

2.2. Introduction of Feed-Forward Neural Network with Radial Basis Function

Two-layer feed-forward neural network as shown in Figure 1 can be presented in matrix form:

$$y = W\sigma(Vx) \tag{4}$$

where $W = [w_{ij}]$ and $V = [v_{jk}]$ are the weight vectors,

$x = \left[1, x_1, \cdots, x_{N_1}\right]^{\mathrm{T}}$ is the input vector,

$y = \left[y_1, y_2, \cdots, y_{N_3}\right]^{\mathrm{T}}$ is the output vector, and

$\sigma = \left[1, \sigma_1, \sigma_2, \cdots, \sigma_{N_2}\right]$ denotes the activation function vector. Constant 1 included in the vector x and σ as a first term allows one to incorporate the thresholds θ_j^v

and θ_i^w

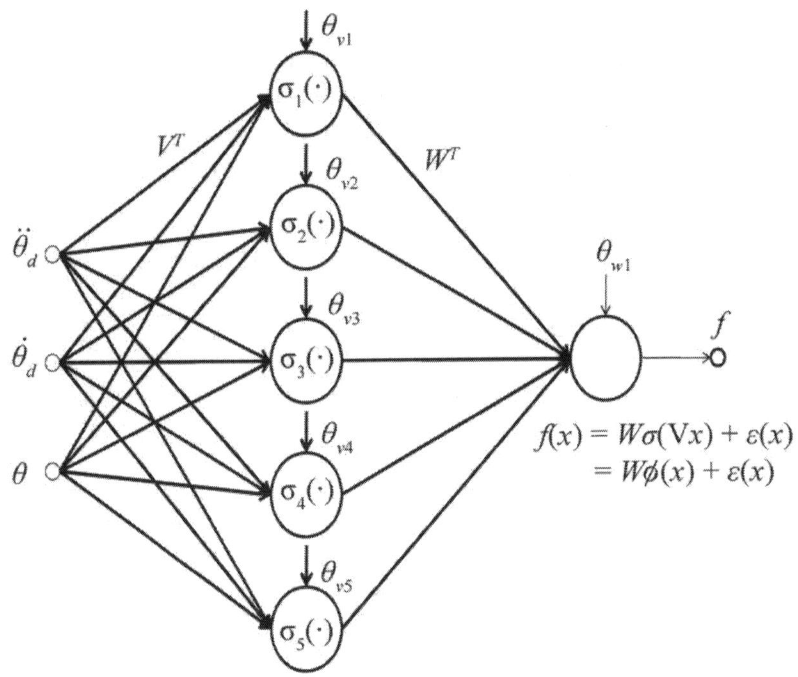

Figure 1. Two-layer feed-forward neural network.

The well-known universal approximation property of a neural network proposed by Hornik et al. [10] says that for a nonlinear function $f(x)$, there exists a sufficiently large N_2 number of neurons satisfying

$$f(x) = W\sigma(Vx) + \varepsilon(x) \tag{5}$$

and makes the two-layer neural network functional approximation error vector $\varepsilon(x)$ arbitrarily small.

We know that the main disadvantage of the multilayer neural network is highly nonlinear in parameter. Hence, we consider hereafter the neural network of fixed V, which makes the neural network linearly parameterized. Defining $\phi(x) = \sigma(Vx)$, we have

$$f(x) = W\phi(x) + \varepsilon(x) \tag{6}$$

where $\phi = \left[\phi_1, \phi_2, \cdots, \phi_{N_2}\right]$ is the generalized basis function vector. In this paper, we use the radial basis function with Gaussian form as following:

$$\phi_i(x) = \exp\left(\frac{\|x - o_i\|}{2\sigma_i^2}\right), i = 1, \cdots, N_2 \tag{7}$$

where o_i is the center of the i-th radial basis function, and σ_i is the width coefficient.

2.3. Controller Design

In this section, we introduce the design of proposed controller for AC motor based on the Lyapunov's stability theorem. The strategy is to define an augment error and design a neural network with learning rules and a robust adaptive term to force the augment error to converge to zero asymptotically in spite of the presence of the system uncertainties and load torque disturbances.

The controller scheme is shown in **Figure 2**.

The angular position tracking error is denoted as:

$$e_\theta = \theta_d - \theta \qquad (8)$$

where θ_d is the desired angular position.

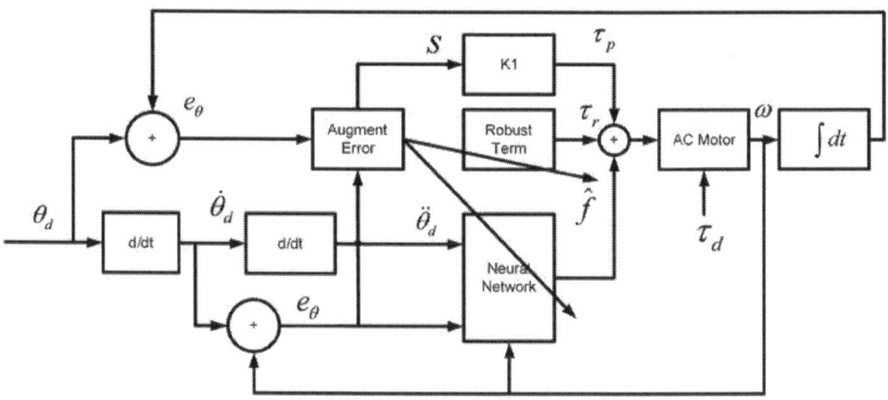

Figure 2. Controller scheme.

Differentiating Equation (8) with respect to time to get the angular position tracking error:

$$e_\omega = \omega_d - \omega \tag{9}$$

The angular velocity tracking error is presented as following:

$$\dot{e}_\omega = \dot{\omega}_d - \dot{\omega} \tag{10}$$

In order to solve the position tracking problem with asymptotical stability, we define the augment error as below:

$$s = e_\omega + \lambda e_\theta \tag{11}$$

Differentiating Equation (11) and substituting it into dynamic Equation (1), we have the error dynamic equation:

$$J\dot{s} = J\ddot{\theta}_d + B\dot{\theta} + J\lambda e_\omega - \tau + \tau_d \tag{12}$$

The nonlinear function containing the dynamic parameters of AC servo motor is defined as:

$$f(x) = J\ddot{\theta}_d + B\dot{\theta} + J\lambda e_\omega \tag{13}$$

It is well-known that the neural network is capable of approximating any nonlinear function over compact input space [10]. The strategy is to construct a simple neural network with radial basis function as presented in Section II.B to approximate the nonlinear function (4). Therefore, the output of neural network controller can be rewritten as:

$$f(x)=W\phi(x)+\varepsilon(x) \tag{14}$$

Assumption 1: There exist unknown positive constants $\varepsilon_{\max}$ and $\tau_{\max}$ such that $\|\varepsilon\| \leq \varepsilon_{\max}$ and $\|\tau\| \leq \tau_{\max}$.

The torque control input is defined as:

$$\tau = \hat{f}(x)+\tau_r+\tau_p \tag{15}$$

$$\hat{f}(x)=\hat{W}\phi(x) \tag{16}$$

$$\tau_r = \hat{\rho}\frac{s}{\|s\|} \tag{17}$$

$$\tau_p = k_1 s \tag{18}$$

where $\hat{f}(x)$ is the estimate function of the nonlinear function $f(x)$, $\hat{W}$ is estimate weight vector of W, τ_r is the robust adaptive control input, τ_p is the stabilizing control term, and k_1 is a constant denoted as the rate of error convergence.

Substituting Equations (14) and (15) into Equation (12), we obtain the closed-loop error dynamics for augment error s:

$$J\dot{s} = \tilde{W}\phi + \varepsilon - \tau_r - \tau_p + \tau_d \tag{19}$$

where $\tilde{W} = W - \hat{W}$ is the estimate weight error.

Consider the Lyapunov function:

$$V = \frac{1}{2}Js^2 + \frac{1}{2k_2}\tilde{W}W^T + \frac{1}{2k_3}\tilde{\rho}^2 \tag{20}$$

Differentiating Equation (20) with respect to time and substituting Equation (19) into the result, we can obtain:

$$\dot{V} = \tilde{W}\left(s\phi + \frac{1}{k_2}\dot{\tilde{W}}^T\right) - k_1 s^2$$
$$-s\tau_r + s\varepsilon + s\tau_d + \frac{1}{k_3}\dot{\tilde{\rho}}\tilde{\rho} \tag{21}$$

By choosing the weight update rule:

$$\dot{\hat{W}} = k_2 s\phi^T \tag{22}$$

and combining with Assumption 1, we can infer the boundedness of Equation (21) as following:

$$\dot{V} \le -k_1 s^2 + \|s\|\rho - \|s\|\hat{\rho} + \frac{1}{k_3}\dot{\tilde{\rho}}\tilde{\rho} \tag{23}$$

where $\rho = \varepsilon_{\max} + \tau_{d\max}$

Since $\tilde{\rho} = \rho - \hat{\rho}$ then

$$\dot{V} \le -k_1 s^2 + \|s\| \tilde{\rho} + \frac{1}{k_3} \dot{\tilde{\rho}} \tilde{\rho} \tag{24}$$

The adaptation rule is simply selected as:

$$\dot{\hat{\rho}} = k_3 \|s\| \tag{25}$$

It can be seen that the adaptation law (25) suppresses the approximation estimation error and disturbance, allows the augment error to converge to zero asymptotically:

$$\dot{V} \le -k_1 s^2 \le 0 \tag{26}$$

Because $s = e_\omega + \lambda e_\theta$ is dynamics stability, $e_\theta = \theta_d - \theta$ and $e_\omega = \omega_d - \omega$ are asymptotically stable.

3. EXPERIMENTAL RESULTS

The proposed motion control board as shown in **Figure 3** is developed based on high performance floating point DSP TMS320C6727 with the CPU clock rate of 350 MHz. Moreover, the proposed controller includes multichannels of 16-bit DAC, DIO, encoder interface, and PCI interface. The servo interface, including encoder module, DAC control module, and DIO are designed in the FPGA XC3S1000. The PCI interface is built in the FPGA to communicate with the host PC. The DSP communicates with the PCI interface through Wishbone Connection Matrix structure. The block diagram of the proposed control board is shown in **Figure 4**.

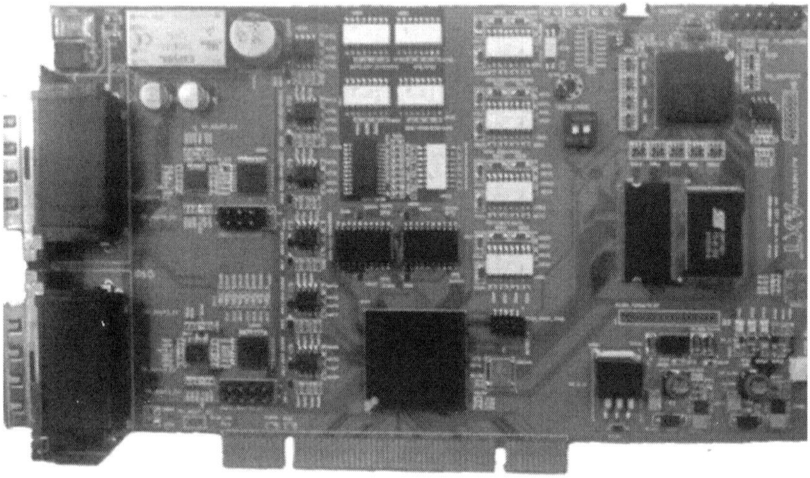

Figure 3. Developed DSP-based motion controller.

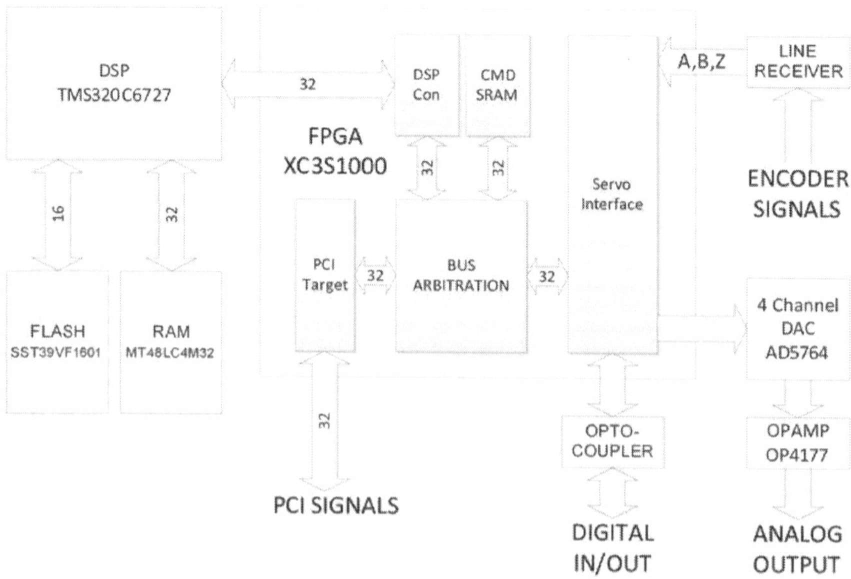

Figure 4. Block diagram of DSP controller.

Code Composer Studio software is used to implement the algorithm in TMS320C6727. The overall algorithm is divided into two main parts: main module and runtime module. The main module defines and initializes the software variables, constants and specific registers. Moreover, some look-up tables employed in the algorithm is also addressed in this parts. Initialized registers in the initialization module are Phase-Locked-Loop (PLL) registers, timer registers, EDMA registers, EMIF registers, interrupt registers, and communication registers. The initialization module is called once at the beginning of the program. Runtime module is generated in the form of functional modules. This modularity technique is preferred to debug and maintain the code in an easy way. In motion control applications, the runtime module including overall control algorithms, trajectory generation, realtime updating of position and velocity variable, and digital filters is carried out by each interrupt clock.

The neural network structure using in the experiment has 3 inputs, 5 neurons in hidden layer, and 1 output layer. The initial weights and thresholds are set as 0.01. The weights vector is updated online by the learning rule proposed in Equation (16).

From the proof of stability, it can be seen that the constant k_1 determines the convergence speed of augment error s. However, the bigger k_1 is, the higher torque is needed to drive the system. k_2 determines the learning speed of the neural network. Therefore, high gain k_2 can avoid the system uncertainties. k_3 decides the adaptive speed of the robust term. Due to the characteristic of dynamics stability of augment error s, the constant λ should be chosen sufficiently small to guarantee the convergence of angular position error. In the simulation, the initial control gains and neural network structure are selected as: $k_1 = 100$, $k_2 = 5$, $k_3 = 100$, $\lambda = 0.01$.

The overall experimental system setup shown in Figure 5 consists of a DSP-based motion controller, an ELMO analog amplifier, a LG FMA-CN02 servo motor with 2000 pulses/revolution feedback encoder, a linear stage RoboStar with 10 mm lead screw, and a host computer.

A circle path with center position (x_c, y_c) and radius R is designed as reference trajectory for the experiment, which can be presented by individual axes as following:

$$x_t = x_c + R\cos\left(\frac{g_t}{R}\right) \tag{27}$$

$$y_t = y_c + R\sin\left(\frac{g_t}{R}\right) \tag{28}$$

The experimental results using conventional PID controller with the control gains $K_P = 15$, $K_I = 200$, $K_D = 0.15$ are shown in Figures 6-12. **Figure 13** shows the circular interpolation tracking of XY table. Figures 7 and 10 are the trajectories of individual axis. Figures 9 and 12 present the input torque command of the PID controller. The tracing error of X-axis and Y-axis in Figures 8 and 11 show that during the motion path, PID controller is not able to reduce the tracking error since it is a linear feedback control system.

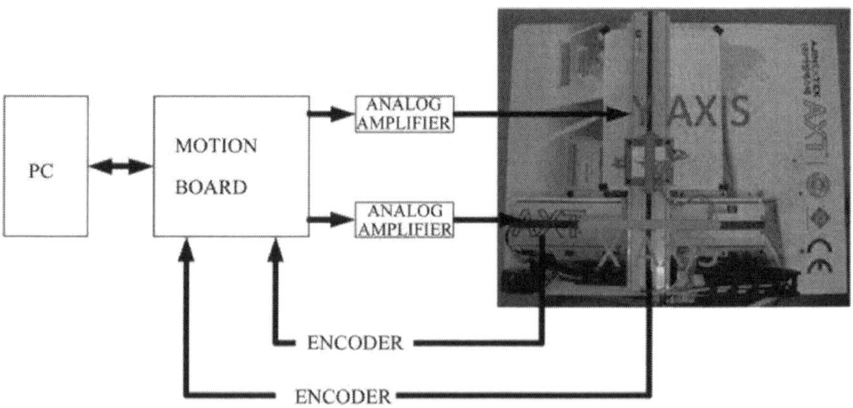

Figure 5. The proposed XY table motion control system.

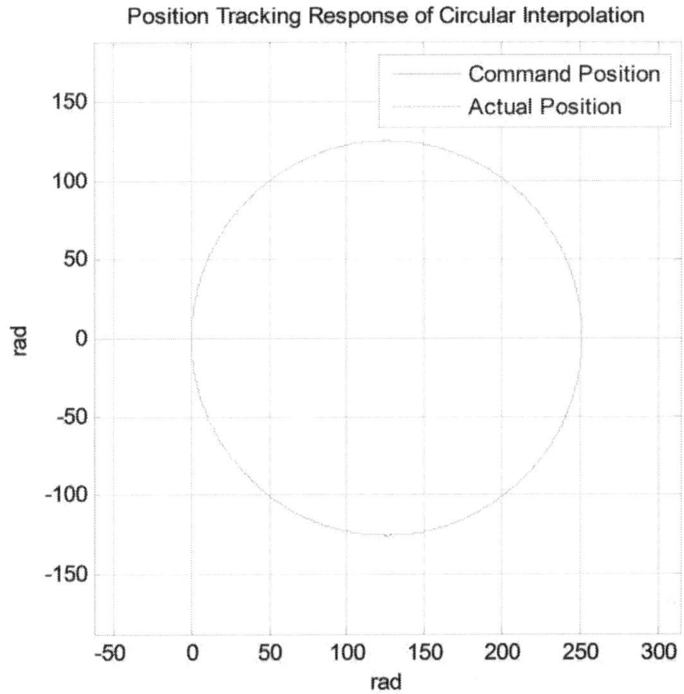

Figure 6. Position tracking response of circular interpolation using PID controller.

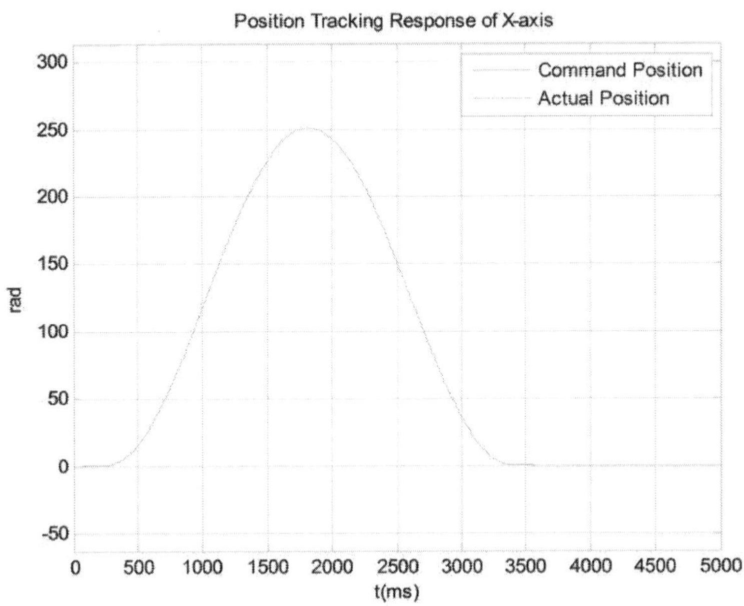

Figure 7. Position tracking response of X-axis using PID controller.

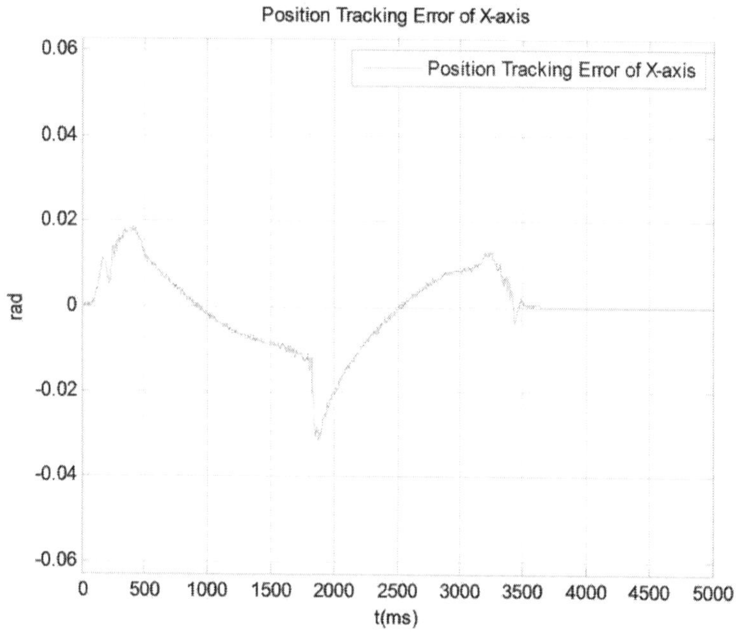

Figure 8. Position tracking error of X-axis using PID controller.

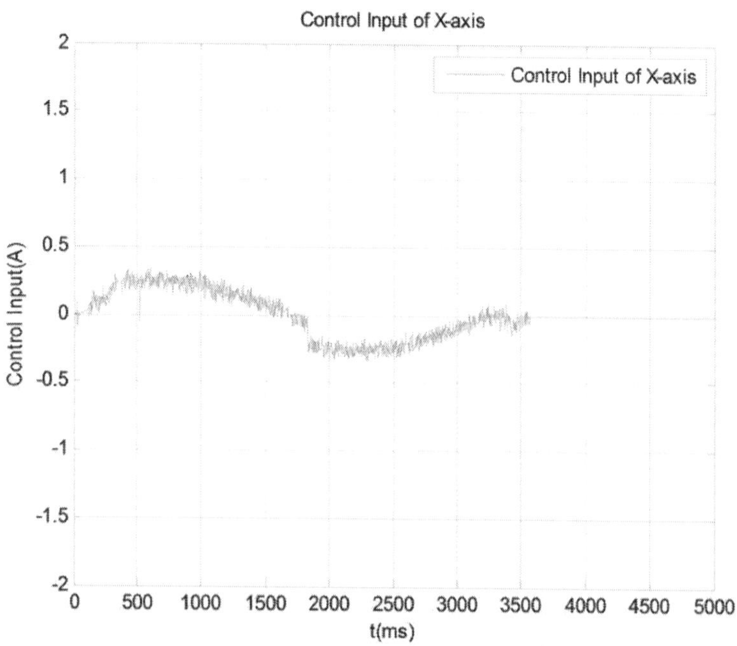

Figure 9. Control input of X-axis using PID controller.

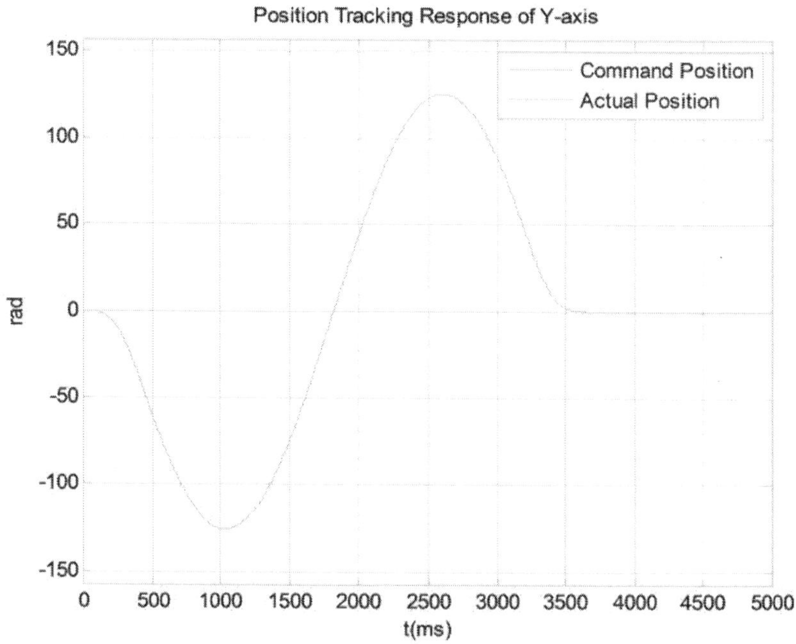

Figure 10. Position tracking response of Y-axis using PID controller.

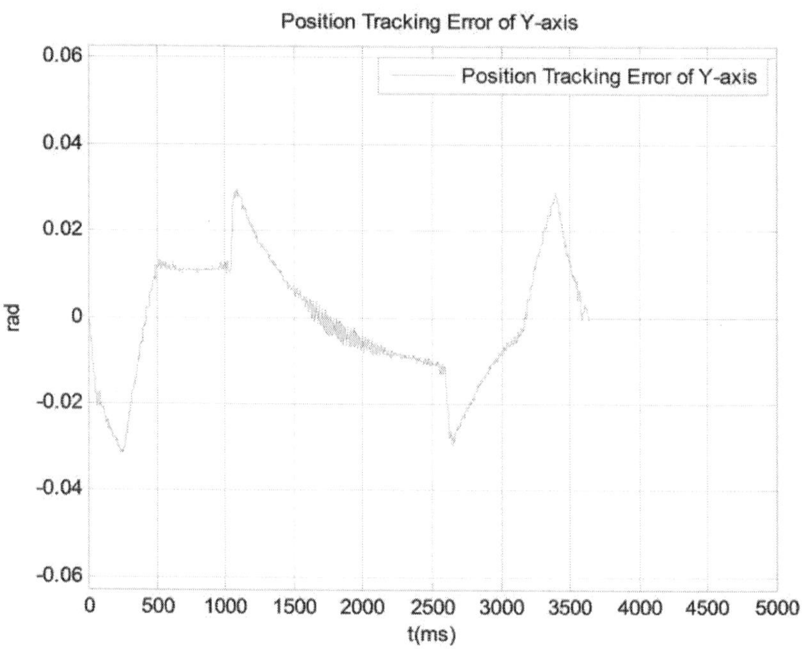

Figure 11. Position tracking error of Y-axis using PID controller.

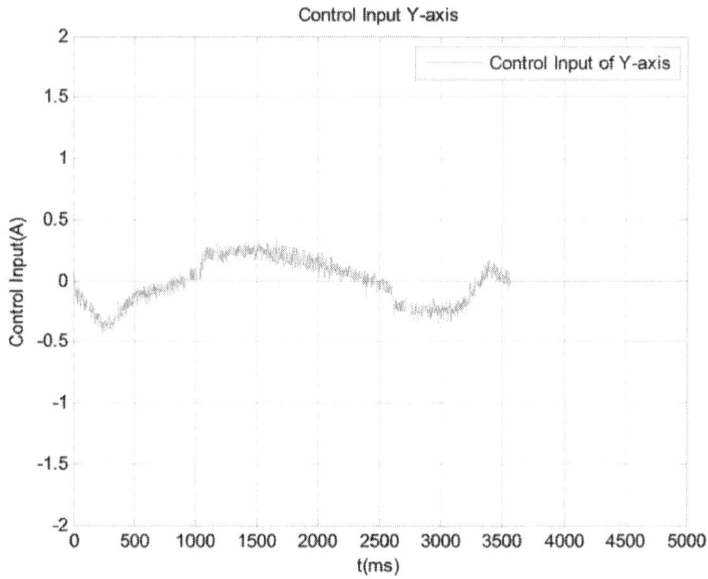

Figure 12. Control input of Y-axis using PID controller.

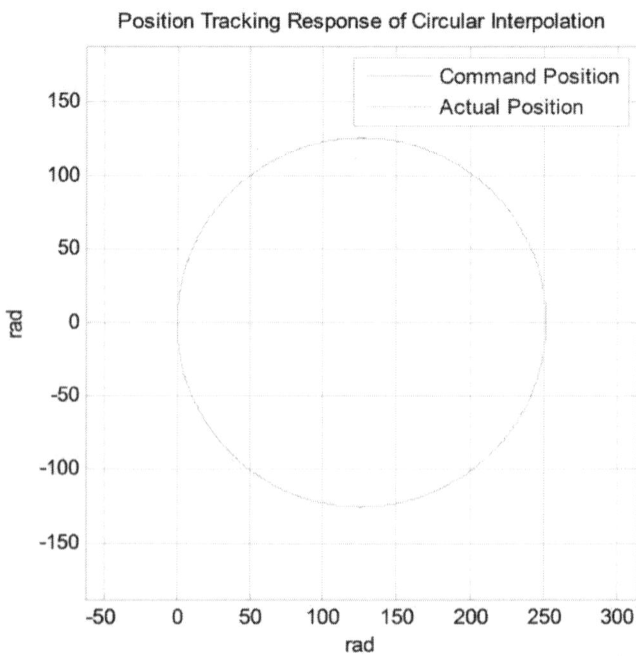

Figure 13. Position tracking response of circular interpolation using RANN controller.

Figures 13-19 show the experimental results using proposed RANN controller. The convergence rate k_1, the neural network learning rate k_2, the adaptive speed k_3, and the constant λ are selected as 120, 0.0005, 100, and 0.01 respectively. The initial weights and thresholds in neural network are set as 0.01. It is clear from the experimental results that the tracking error is substantially reduced to nearly zero. The actual positions of the motors closely track the reference and tracking error converges to zero quickly. Moreover, the control input responses of X-axis and Y-axis shown in Figures 16 and 19 indicate the improvement of control effort of proposed controller, since it keeps the tracking error negligible during the motion path. On the contrary, the PID controller cannot produce sufficient torque command to keep tracking error of motor arbitrary small. **Table 1** shows the comparison of the dynamic performance of the PID controller and RANN controller. It can be seen from the **Table 1** that the proposed controller has the standout performance over the traditional PID controller.

From the experimental results, it is clear that by using the proposed controller, the convergence and stability of the servo system can be guaranteed despite the existence of model uncertainties and disturbances and the rotor angle follows its desired trajectory satisfactorily.

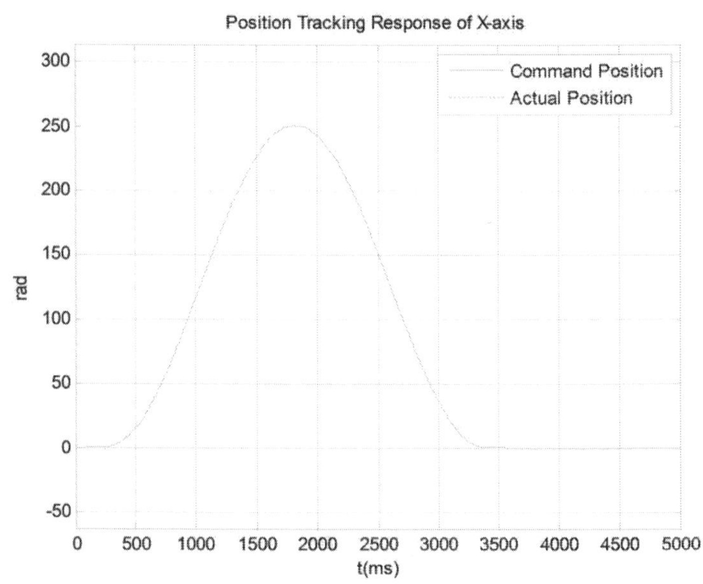

Figure 14. Position tracking response of X-axis using RANN controller.

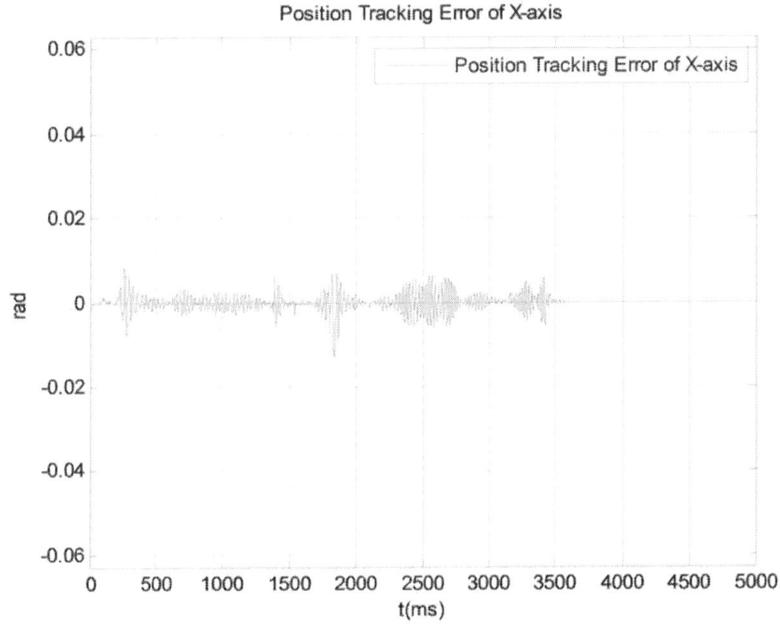

Figure 15. Position tracking error of X-axis using RANN controller.

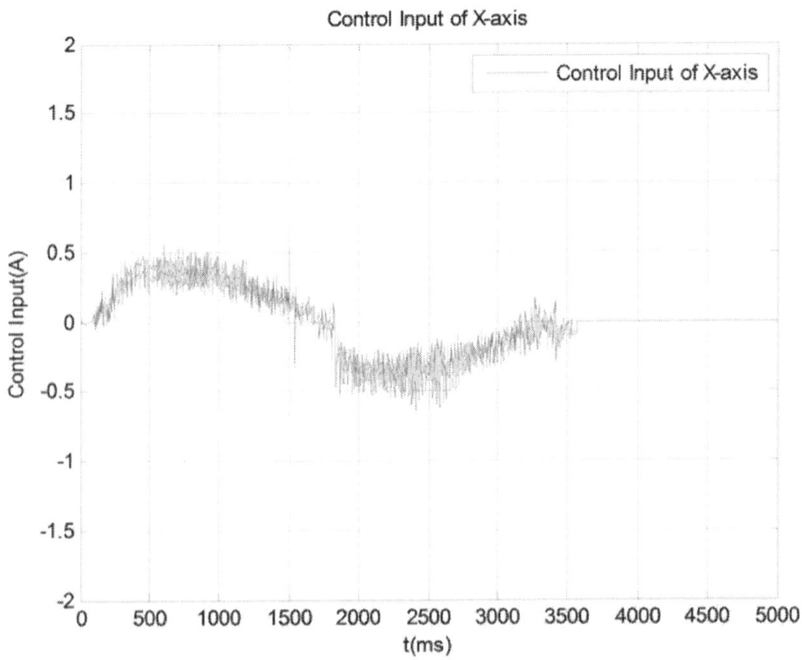

Figure 16. Control input of X-axis using RANN controller.

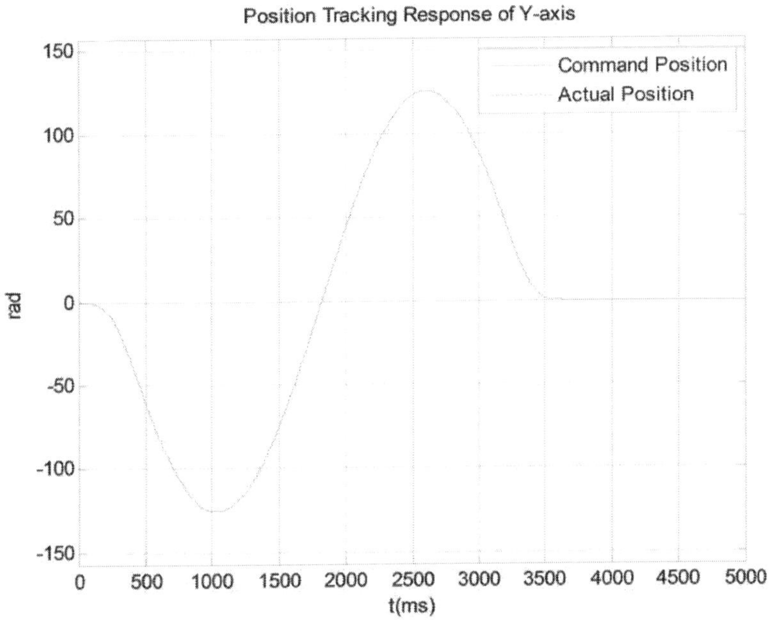

Figure 17. Position tracking response of Y-axis using RANN controller.

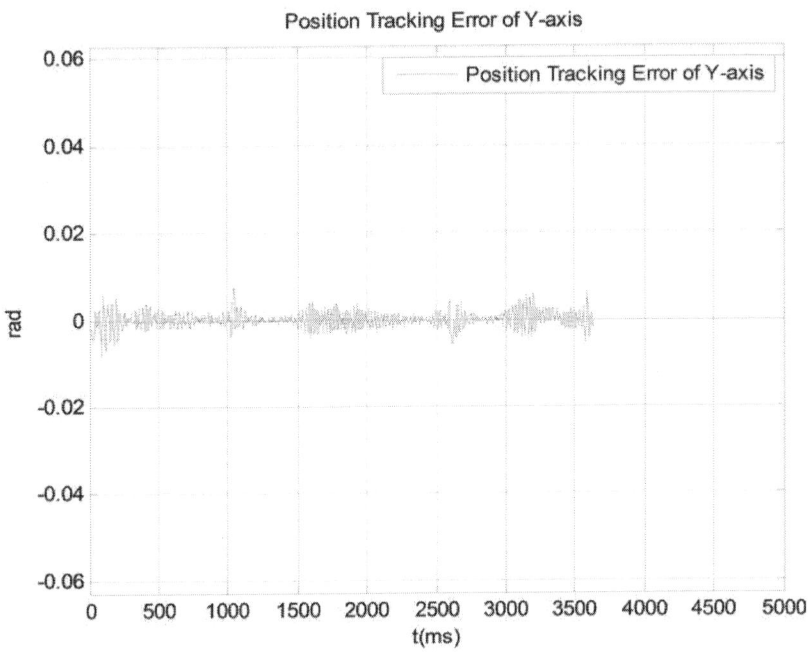

Figure 18. Position tracking error of Y-axis using RANN controller.

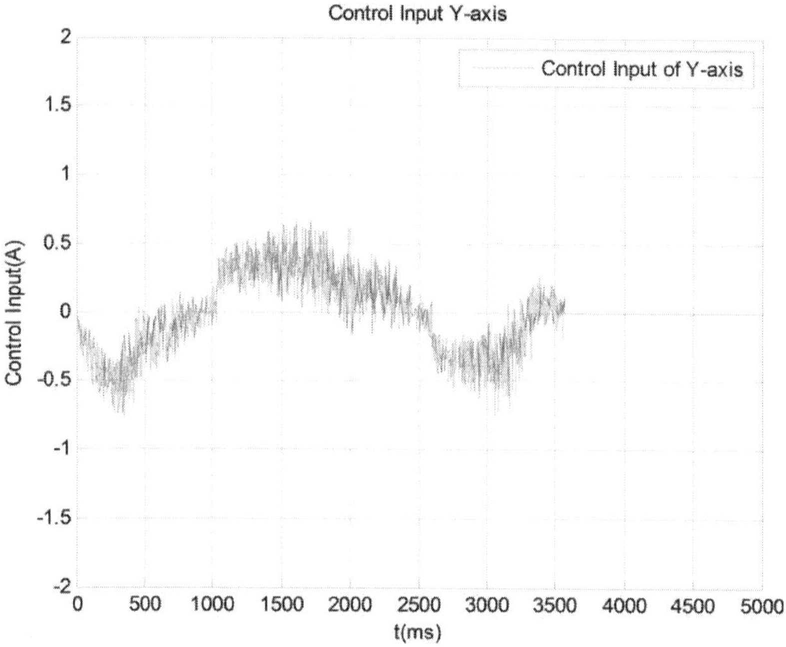

Figure 19. Control input of Y-axis using RANN controller.

Table 1. Comparison of PID and RANN controller.

	Controller	
	PID	RANN
Max error (rad)	0.0341	0.0133
Mean error (rad)	0.0118	0.0017
RMS error (rad)	0.0150	0.0024

4. CONCLUSIONS

In this paper, a robust adaptive neural network controller has been successfully implemented for the X-Y table.

The proposed schemes are designed by a systematic approach and implemented based on the high performance DSP TMS320C6727. Fast learning rule of the neural network and stable adaptive mechanism enable the controlller to overcome the nonlinear friction, uncertainties and disturbances of the mechanical system.

The stability and the convergence of the system are proved by Lyapunov theory and the validity of the controller is verified through experimental results.

REFERENCES

1 H. Lim, J. W. Seo and C. H. Choi, "Position Control of XY Table in CNC Machining Center with Non-Rigid Ballscrew," Proceedings of the American Control Conference, Chicago, 2000, pp. 1542-1546.

2 E. C. Park, H. Lim and C. H. Choi, "Position Control of X-Y Table at Velocity Reversal Using Presliding Friction Characteristics," IEEE Transactions on Control Systems Technology, Vol. 11, No. 1, 2003, pp. 24-31.

3 F. J. Lin and P. H. Shen, "Robust Fuzzy Neural Network Sliding-Mode Control for Two-Axis Motion Control System," IEEE Transactions on Industrial Electronics, Vol. 53, No. 4, 2006, pp. 1209-1225.

4 Z. Jamaludin, H. Van Brussel and J. Swevers, "Friction Compensation of an XY Feed Table Using Friction-Model-Based Feedforward and an Inverse-Model-Based Disturbance Observer," IEEE Transactions on Industrial Electronics, Vol. 56, No. 10, 2009, pp. 3848-3853.

5 Y. T. Kim, "Adaptive Fuzzy Backstepping Control of AC Servo System in the Presence of Nonlinear Dynamic Effect and Mechanical Uncertainties," Automation Congress, WAC, Hawaii, 28 September-2 October 2008, pp. 1-8.

6 J. Chang, Y. Tan and J. T. Yu, "Backstepping Approach of Adaptive Control, Gain Selection and DSP Implementation for AC Servo System," IEEE Power Electronics Specialists Conference, Orlando, 17-21 June 2007, pp. 535- 541.

7 Y. S. Xiao, Q. D. Wu and G. X. Zhou, "Neural Network Based Parameters Identification and Adaptive Speed Control of AC Drive System," Proceedings of the IEEE International Conference on in Industrial Technology, Shanghai, 2-6 December 1996, pp. 118-121.

8 Y. X. Su, B. Y. Duan and Y. F. Zhang, "Robust Precision Motion Control for AC Servo System," Proceedings of the 4th World Congress on Intelligent Control and Automation, Vol. 4, 2002, pp. 3319-3323.

9 P. H. Kim, S. H. Sin, H. L. Baek and G. B. Cho, "Speed Control of AC Servo Motor Using Neural Networks," Proceedings of the 5th International Conference on Electrical Machines and Systems, Shenyang, August 2001, pp. 691-694.

10 K. Hornik, M. Stinchcombe and H. White, "Multilayer Feed-Forward Networks Are Universal Approximator," Neural Networks, Vol. 2, No. 5, 1989, pp. 359-366.

CHAPTER 7

Robust Adaptive Control for a Class of Systems with Deadzone Nonlinearity

Nizar J. Ahmad*, **Mahmud J. Alnaser, Ebraheem Sultan, Khuloud A. Alhendi**

Faculty of Electronic Engineering Technology, College of Technological Studies, The Public Authority for Applied Education and Training (PAAET), Kuwait City, Kuwait

ABSTRACT

This paper presents a robust adaptive control scheme for a class of continuous-time linear systems with unknown non-smooth asymmetrical deadzone nonlinearity at the input of the plant. The methodology is applied to handle input deadzone as well as unmeasurable disturbances si- multaneously in strictly matched systems. The proposed controller robustly cancels any residual distortion caused by the inaccurate deadzone cancellation scheme. The scheme is shown to suc- cessfully cancel the deadzone's deleterious effect as well as eliminate other unmeasurable distur- bances within the span of the input. The new controller ensures the global stability of all states and adaptations, and achieves asymptotic tracking. The asymptotic stability of the closed-loop system is proven by Lyapunov arguments, and simulation results confirm the efficacy of the con- trol methodology.

KEYWORDS

Adaptive Control, Non-Symmetric Deadzone, Hard Nonlinearity

1. INTRODUCTION

The significance of the deadzone problem lies in the fact that it affects many physical and practical systems. Examples of such systems are the ones containing hydraulic or pneumatic values, electronic circuits and devices, temperature regulation circuits, and in actuators such as servo valves and DC motors. In most cases, the pa- rameters of the deadzone nonlinearity are unknown and continuously varying with time and temperature. Deadzones exist in a number of industrial applications specially the ones requiring high precision such as medical robots, semiconductor manufacturing, and precision machine tools. It has been shown that deadzone in actuators, such as hydraulic servo-valves, gives rise to limit cycling and instability. The advances reached in the area of adaptive compensation and control theory gave rise to increased interest in handling the deadzone problem. There have been many techniques that addressed the problem and have been shown to reduce if not eliminate the degradation of system performance resulting in an improved tracking accuracy and ensured stabilization of such systems.

One sensible approach to counter the effect of the deadzone, shown in Figure 1, was presented by [1] which involved designing an inverse deadzone function to cancel its effect. The approach of designing an inverse adaptive deadzone compensator was thoroughly investigated in [2] and [3] which was shown to improve per- formance. Lewis et al. in [4] proposed a fuzzy logic type inverse deadzone compensator, meanwhile, a neural network inverse compensator was designed in [5] . Both approaches show clear improvement in reducing the tracking error. In [6] , a new adaptive controller of linear or nonlinear systems with deadzone is introduced with- out constructing a deadzone inverse. Global and asymptotic tracking was achieved and simulation results were presented.

An adaptive sliding mode control scheme used to offset a non-symmetrical deadzone nonlinearity in continu- ous time was presented in [7]. The problem of chattering inherent with sliding mode control is handled by al- lowing a small controlled tracking error.

In recent years, many researchers addressed the deadzone problem with encouraging results. In [8] , a novel function was introduced to describe deadzone nonlinearity. To show the effectiveness of the proposed equivalent function the authors combined it with vibration of a cantilever beam.

In this paper we are motivated by the success of our earlier results deadzone compensation of DC motor pre- sented in [9] and [10] . The extension involves combining the deadzone for a class of linear systems with uncer- tainties in the span of the input. The uncertainties are assumed to be bounded by a pth order polynomial in the state of the system. A robust adaptive controller will compensate for the unmeasurable disturbances as well as any mismatch error in estimating the deadzone parameters. The proposed method does not require any know- ledge of the deadzone parameters or the specialized design of an inverse deadzone controller and only an upper bound of the deadzone spacing which is easily determined a priori.

2. THE PROBLEM SETUP: DYNAMICS OF A NON-SYMMETRICAL DEADZONE NONLINEARITY

A common representation of a non-symmetrical deadzone nonlinearity, shown in Figure 1, described in [1] as follows

$$DZ\big(u(t)\big) = \begin{pmatrix} m(u-d_r), & \text{if} \quad u > d_r \\ 0, & \text{if} \quad -d_l < u < d_r \\ m(u+d_l), & \text{if} \quad u < -d_l \end{pmatrix} \qquad (1)$$

where $DZ(u(t))$ denotes the output of deadzone function preceding a plant input, m is the slope of the lines, $(d_r - d_l)$ is the width of the deadzone distance, and $u(t)$ is the input of the deadzone block as shown in Figure 1. A more convenient representation of a non-symmetrical deadzone was presented in [6] as

$$DZ(u) = u - sat_d(u) \tag{2}$$

where $sat_d(u)$ represents a non-symmetrical saturation function given by

$$sat_d(u) = \begin{pmatrix} d_r & \text{if} & u > d_r \\ u, & \text{if} & -d_l < u < d_r \\ -d_l, & \text{if} & u < -d_l \end{pmatrix} \tag{3}$$

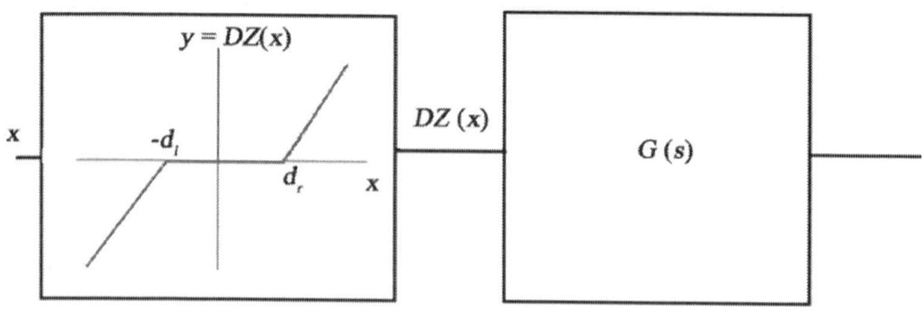

Figure 1. Non-symmetric deadzone nonlinearity at the input of a linear plant.

A novel representation for (2) was presented in [11] for an exact symmetrical deadzone by defining $(d = d_r - d_l)$ as the deadzone spacing function as

$$DZ(u) = m/2(2u + |u - d| - |u + d|) \tag{4}$$

Consequently, the saturation function can be written as

$$sat_d(u) = m/2(|u-d|-|u+d|) \tag{5}$$

In most applications the deadzone parameters are unknown or time and temperature varying. Instead, for de- veloping an inverse deadzone function as in [9] , we advocate a robust adaptive compensator that handles the variability of the deadzone parameters as part of an input disturbance. However, we outline some necessary and reasonable assumptions to be used in the proof of the efficacy of the proposed control methodology:

(A1) The deadzone parameters $d_r > 0$ and $d_l > 0$.

(A2) The deadzone parameters d_r, and d_l are bounded as follows:

$d_l \in [d_l\text{min}, d_l\text{max}]$ and.

(A3) Without any loss of generality, the slope of the deadzone m is positive and is set to 1.

(A4) The output of the deadzone block $DZ(u)$ is not available for measurement.

Remark 1. Assumptions (A1) and (A2) are the actual physical attributes of a real industrial deadzone and is adopted in the literature [6] . Therefore, the saturation function given by (4) can be shown to have an upper bound by closely analysing Equation (5). Using the general inequality rule $|a \pm b| < |a| + |b|$ we have

$$\|u-d\| \le \|u\| + \|d\| \tag{6}$$

$$\|u+d\| \le \|u\| + \|d\| \tag{7}$$

From Equations (6) from (7) we can state

$$\left(\|u-d\|-\|u+d\|\right)\leq 0 \tag{8}$$

Multiplying by the slope $\frac{m}{2}$ gives

$$\frac{m}{2}\left(\|u-d\|-\|u+d\|\right)\leq 0 \tag{9}$$

The left hand side of (9) is the saturation function given by (5). In case of a non-symmetrical deadzone func- tion the upper-bound may be chosen by employing assumption (A2) as follows

$$sat_d\left(u\right)\leq d_M \tag{10}$$

Where

$$d_M=\frac{m}{2}\left(\|d_r\max\|-\|d_l\min\|\right)\geq 0 \tag{11}$$

The upper bounds will play a pivotal role to ensure the overall global stability of the close loop dynamics as will be demonstrated in the following section.

3. ROBUST ADAPTIVE CONTROLLER DESIGN

Considering the following nonlinear system with input deadzone nonlinearity described as

$$\begin{aligned}\dot{x}&=Ax+B\left\{DZ\left(u\right)+\psi\left(x\right)\right\}\\ y&=Cx,\end{aligned} \tag{12}$$

where the matrices A and B are given by

$$A = \begin{pmatrix} 0 & 1 & \cdots & 0 \\ 0 & 0 & 1 & 0 \\ \vdots & \vdots & \ddots & \vdots \\ 0 & 0 & \cdots & 0 \end{pmatrix}, \quad B = \begin{pmatrix} 0 \\ \vdots \\ 0 \\ 1 \end{pmatrix}$$

And $\psi(x)$ represents the unmeasurable disturbance. The collective bounds can be expressed as

$$\|\psi(x)\| \leq \sum_{k=0}^{p} \gamma_k \|x\|^k \tag{13}$$

Let the reference model to be tracked given by

$$\dot{x}_m = Ax_m + B\{Kx_m + r\} \tag{14}$$

where $K \in R^{1xn}$ and r is a reference signal. The tracking error dynamics $\tilde{x} = x - x_m$ may be written as fol- lows:

$$\dot{\tilde{x}} = A\tilde{x} + B\{DZ(u) + \psi(x) - Kx_m - r\} \tag{15}$$

Inserting the deadzone equation (2) into (15) yields

$$\dot{\tilde{x}} = A\tilde{x} + B\{u - sat(u) + \psi(\tilde{x}) - Kx_m - r\} \tag{16}$$

Therefore, for the class of systems described in (13) and deadzone given in (7), we use the result stated as Lemma RANDM in [12] and modify it to ensure asymptotic convergence. The modified controller is

$$u_d(t) = +Kx_m + r - \alpha B^T P\tilde{x} - \hat{\beta} B^T P\tilde{x} + \rho \tanh\left[(a+bt) B^T P\tilde{x}\right] \tag{17}$$

where α, a, $b > 0$, $\rho \geq d_M$, and P is the positive definite symmetric solution of the Algebraic Riccati equation (ARE). The adaptation $\hat{\beta}$ is used to ensure robustness of the controller. The combined output of the com- pensator and the deadzone nonlinearity may be written as

$$DZ(u_d) = u_d - sat(u_d) = sat(u_d) + Kx_m + r - \alpha B^T P\tilde{x} - \hat{\beta} B^T P\tilde{x} + \rho \tanh\left[(a+bt) B^T P\tilde{x}\right] \tag{18}$$

Inserting the proposed control laws (17) and the output of the deadzone block given by Equation (18) into the error dynamics (16) results in the closed loop dynamics

$$\dot{\tilde{x}} = A\tilde{x} + B\left\{-sat(u_d) - \alpha B^T P\tilde{x} - \hat{\beta} B^T P\tilde{x} + \rho \tanh\left[(a+bt) B^T P\tilde{x}\right] + \psi(\tilde{x})\right\} \tag{19}$$

The adaptation law for $\hat{\beta}$ given by

$$\dot{\hat{\beta}} = \Gamma B^{\mathrm{T}} P \tilde{x}^2 \tag{20}$$

where $\Gamma > 0$ is a constant scalar.

Theorem. For the plant described by (12) with input deadzone (1), and the robust adaptive control law (17) along with the adaptive update law (20) will ensure the closed-loop stability and boundedness of tracking error, hence reducing the effects of deadzone.

Proof. Using the following positive definite control Lyapunov function

$$V = \tilde{x}^{\mathrm{T}} P \tilde{x} + \frac{\Gamma^{-1}}{2} \tilde{\beta}^2 \tag{21}$$

where $\hat{\beta} = \tilde{\beta} + B^*$ differentiating along the trajectories of the system yields

$$\dot{V} = \dot{\tilde{x}}^{\mathrm{T}} P \tilde{x} + \tilde{x}^{\mathrm{T}} P \dot{\tilde{x}} + \Gamma^{-1} \tilde{\beta} \dot{\hat{\beta}} \tag{22}$$

$$= \left(A\tilde{x} + B\left\{ DZ(u) + \psi(x) \right\} \right)^{\mathrm{T}} P\tilde{x} + \tilde{x}^{\mathrm{T}} P \left(A\tilde{x} + B\left\{ DZ(u) + \psi(x) \right\} \right) + \Gamma^{-1} \tilde{\beta} \dot{\hat{\beta}} \tag{23}$$

Substituting for the closed loop dynamics given by (19) in (23) gives

$$\begin{aligned} \dot{V} = {} & \left(A\tilde{x} + B\left\{ -sat(u_d) - \alpha B^{\mathrm{T}} P\tilde{x} - \hat{\beta} B^{\mathrm{T}} P\tilde{x} + \rho \tanh\left[(a+bt) B^{\mathrm{T}} P\tilde{x} \right] + \psi(x) \right\} \right)^{\mathrm{T}} P\tilde{x} \\ & + \tilde{x}^{\mathrm{T}} P \left(A\tilde{x} + B\left\{ -sat(u_d) - \alpha B^{\mathrm{T}} P\tilde{x} - \hat{\beta} B^{\mathrm{T}} P\tilde{x} + \rho \tanh\left[(a+bt) B^{\mathrm{T}} P\tilde{x} \right] + \psi(x) \right\} \right) + \Gamma^{-1} \tilde{\beta} \dot{\hat{\beta}}. \end{aligned} \tag{24}$$

Collecting terms and simplifying

$$\dot{V} = \tilde{x}^{\mathrm{T}}\left(A^{\mathrm{T}}P + PA\right)\tilde{x} + 2\tilde{x}^{\mathrm{T}}PB\left\{-\alpha B^{\mathrm{T}}P\tilde{x} - \hat{\beta}B^{\mathrm{T}}P\tilde{x} + \psi(x)\right\}$$
$$+2\tilde{x}^{\mathrm{T}}PB\left\{-sat(u_d) + \rho\tanh\left[(a+bt)B^{\mathrm{T}}P\tilde{x}\right]\right\} + \Gamma^{-1}\tilde{\beta}\dot{\hat{\beta}}. \quad (25)$$

Rearranging terms we get

$$\dot{V} = \tilde{x}^{\mathrm{T}}\left(A^{\mathrm{T}}P + PA - 2\alpha PBB^{\mathrm{T}}P\right)\tilde{x} - \hat{\beta}B^{\mathrm{T}}P\tilde{x}\tilde{x}^{\mathrm{T}}PB + 2\tilde{x}^{\mathrm{T}}PB\psi(x)$$
$$+2\tilde{x}^{\mathrm{T}}PB\left\{-sat(u_d) + \rho\tanh\left[(a+bt)B^{\mathrm{T}}P\tilde{x}\right]\right\} + \Gamma^{-1}\tilde{\beta}\dot{\hat{\beta}}. \quad (26)$$

The first term can be simplified by solving the Algebraic Reccati Equation given by

$$A^{\mathrm{T}}P + PA - 2\alpha PBB^{\mathrm{T}}P = -Q \quad (27)$$

resulting in

$$\dot{V} = -\tilde{x}^{\mathrm{T}}Q\tilde{x} - \hat{\beta}\tilde{x}^{\mathrm{T}}PB^{2} + 2\tilde{x}^{\mathrm{T}}PB\ \psi(\tilde{x}) + \Gamma^{-1}\tilde{\beta}\dot{\hat{\beta}} - 2\tilde{x}^{\mathrm{T}}PB\left(sat(u_d) - \rho\tanh\left[(a+bt)B^{\mathrm{T}}P\tilde{x}\right]\right) \quad (28)$$

Replacing the adaptation law (20) and replacing $\hat{\beta} = \tilde{\beta} + \beta^{*}$ in (28) yields

$$\dot{V} = -\tilde{x}^{\mathrm{T}} Q \tilde{x} - \left(\tilde{\beta} + \beta^*\right) \tilde{x}^{\mathrm{T}} PB^2 + \tilde{\beta} \tilde{x}^{\mathrm{T}} PB^2 + 2\tilde{x}^{\mathrm{T}} PB \psi(x) - 2\tilde{x}^{\mathrm{T}} PB\left(sat(u_d) - \rho \tanh\left[(a+bt) B^{\mathrm{T}} P\tilde{x}\right]\right) \tag{29}$$

$$\dot{V} = -\tilde{x}^{\mathrm{T}} Q \tilde{x} - \beta^* \tilde{x}^{\mathrm{T}} PB^2 + 2\tilde{x}^{\mathrm{T}} PB\left[\left(\psi(x)\right)\right] - 2\tilde{x}^{\mathrm{T}} PB\left(sat(u_d) - \rho \tanh\left[(a+bt) B^{\mathrm{T}} P\tilde{x}\right]\right) \tag{30}$$

So far the first two terms are negative. As for the third term we utilize the general inequality $2ab \le a^2 + b^2$ to establish proper bounds as follows

$$\left\|2\tilde{x}^{\mathrm{T}} PB \cdot \psi(x)\right\| \le \varsigma \left\|\tilde{x}^{\mathrm{T}} PB\right\|^2 + \varsigma^{-1} \left\|\psi(x)\right\|^2 \tag{31}$$

Using the inequality (13) to modify (31) to become

$$\varsigma \left\|\tilde{x}^{\mathrm{T}} PB\right\|^2 + \varsigma^{-1} \left\|\psi(x)\right\|^2 \le \varsigma \left\|\tilde{x}^{\mathrm{T}} PB\right\|^2 + \varsigma^{-1} \gamma \left\|\tilde{x}\right\|^2 \tag{32}$$

Therefore, the inequality of (32) can be incorporated in (30) as

$$\dot{V} \le -\left(\lambda_{\min}(Q) - \varsigma^{-1}\gamma\right)\left\|\tilde{x}\right\| - \beta^* \left\|\tilde{x}^{\mathrm{T}} PB\right\|^2 + \varsigma \left\|\tilde{x}^{\mathrm{T}} PB\right\|^2 - 2\tilde{x}^{\mathrm{T}} PB\left(sat(u_d) - \rho \tanh\left[(a+bt) B^{\mathrm{T}} P\tilde{x}\right]\right) \tag{33}$$

By choosing the degree of freedom ς satisfying the condition $\varsigma < \frac{\lambda_{\min}(Q)}{\gamma}$ and choosing β^* to be greater than ς ensures that the first three terms of $\dot{V}$ negative. Meanwhile, the last term in (33) can be upper bounded as follow

$$-2\tilde{x}^{\mathrm{T}}PB\left(sat\left(u_{d}\right)-\rho\tanh\left\{\left(a+bt\right)B^{\mathrm{T}}P\tilde{x}\right\}\right)\leq-2\left\|\tilde{x}^{\mathrm{T}}PB\right\|\left(\left\|sat\left(u_{d}\right)\right\|-\rho\left(\tanh\left\{\left(a+bt\right)\left\|\tilde{x}^{\mathrm{T}}PB\right\|\right\}\right)\right) \tag{34}$$

Utilizing the upper bounds on $\left\|sat\left(u_{d}\right)\right\|<d_{M}$ given by (10) and rewriting the right hand side of (34)

$$-2\left\|\tilde{x}^{\mathrm{T}}PB\right\|\cdot\rho\cdot\left(\delta-\tanh\left[\left(a+bt\right)\left\|\tilde{x}^{\mathrm{T}}PB\right\|\right]\right) \tag{35}$$

where $\delta=d_{M}/\rho<1$. Therefore the last term in (35) insures that $\dot{V}\leq 0$ as long as

$$\delta-\tanh\left[\left(a+bt\right)\left\|\tilde{x}^{\mathrm{T}}PB\right\|\right]>0 \tag{36}$$

Or

$$\left\|\tilde{x}^{\mathrm{T}}PB\right\|>\Pi(t)\square\frac{1}{2\left(a+bt\right)}\ln\left(\frac{1+\delta}{1-\delta}\right) \tag{37}$$

To conclude, by choosing $\rho>d_{M}$ then $\dot{V}$ is rendered negative and $\left\|\tilde{x}^{\mathrm{T}}PB\right\|$ converges to a closed and vanishing region as time increases. Therefore, since $\left\|\tilde{x}^{\mathrm{T}}PB\right\|\to 0$ as $t\to 0$ and $\tilde{x}^{\mathrm{T}}PB=c_{1}\tilde{x}_{1}+c_{2}\tilde{x}_{2}+\cdots+c_{n}\tilde{x}_{n}=0$

implies that by choosing the c vector as the coefficient of a strictly Hurwitz polynomial will make the closed loop system error asymptotically stable. For a more through conclusion of the proof one may refer to [13] .

4. ILLUSTRATIVE EXAMPLE

In this section, we illustrate the proposed controller to compensate for a system with a deadzone nonlinearity presented in [14] as

$$\ddot{x} = a_1 \frac{(1-e^{-x})}{(1+e^{-x})} - a_2(\dot{x}^2 + 2x)\sin(\dot{x}) - 0.5a_3 x\sin(3t) + DZ(u) \tag{38}$$

The parameter used for the simulation is shown in Table 1. The plant (41) may be written in state space form by defining $x = x_1$ and $\dot{x} = x_2$ then

$$\dot{x}_1 = x_2 \tag{39}$$

$$\dot{x}_2 = a_1 \frac{(1-e^{-x_1})}{(1+e^{-x_1})} - a_2(x_2^2 + 2x_1)\sin(x_2) - 0.5a_3 x_1\sin(3t) + DZ(u)$$

Resulting in

$$A = \begin{pmatrix} 0 & 1 \\ 0 & 0 \end{pmatrix}, \quad B = \begin{pmatrix} 0 \\ 1 \end{pmatrix}$$

Table 1. Parameters utilized in the example.

Item	Systems Physical Attributes		
	Parameter	Value	Unit
1	k_p	40.0	PD Gain
2	k_v	13.0	PD gain
3	d_r^*	20.0	Deadzone Right
4	d_l^*	15.0	Deadzone Left
6	Γ	10.0	Gains
7	$b,\ a_1,\ a_2,\ a_3$	1.0	Scalars
9	ρ	4.0	Gain
10	α	0.2	Gain
11	$a,\ b$	0.2	Scalars

The solution of the ARE equation was chosen to be

$$\boldsymbol{P} = \begin{pmatrix} 130.17 & 22.36 \\ 22.36 & 14.55 \end{pmatrix}, \quad \boldsymbol{Q} = \begin{pmatrix} 200 & 0 \\ 0 & 40 \end{pmatrix}$$

The reference model to be tracked is

$$\dot{x}_m = \begin{pmatrix} 0 & 1 \\ -k_p & -k_v \end{pmatrix} \begin{pmatrix} x_{m1} \\ x_{m2} \end{pmatrix} + \begin{pmatrix} 0 \\ 1 \end{pmatrix} r(t)$$

for a sinusoidal reference trajectory given by

$$r(t) = 2 + \sin(\omega_d t) + \frac{1}{2}\sin(5\omega_d t)$$

Meanwhile, the unmeasurable disturbance $\psi(x)$ can be collectively bounded as

$$\begin{aligned}\|\psi(x)\| &= a_1\frac{\left(1-e^{-x_1}\right)}{\left(1+e^{-x_1}\right)} - a_2\left(x_2^2+2x_1\right)\sin\left(x_2\right) - 0.5a_3x_1\sin\left(3t\right)\\ &\leq \sum_{k=0}^{p}\gamma_k\|x\|^k\end{aligned} \tag{40}$$

Simulations of the system in (39) under the adaptive control law (17) and (20) have been performed. The up- per bounds on actuator actual spacing $d_M = d_r - d_l = 35$ is assumed unknown. In order to demonstrate the performance improvement accomplished by our proposed method, the system under test given in (39) was used. The efficacy of the proposed method is proven by comparing its performance against the performance of a clas- sic PD controller having equivalent gains. The complete parameters of the system under test and controller gains are listed in Table 1.

The simulation results presented in Figure 2 and Figure 3, clearly show the tracking performance for x_1 and x_2 states along with their respective reference trajectory. Figure 4 and Figure 5demonstrate the tracking error for the states $\tilde{x}_1$ and $\tilde{x}_2$ in blue in addition to the same tracking errors for the system under a PD con- troller in red. In both figures, the PD controller resulted in limit cycles where as the adaptive controller proved to be stable with no limit cycles and improved performance with a zero approaching tracking error. Figure 6 shows the control effort $u_d(t)$ and $DZ(u_d)$. In Figure 7 the evolution of the $\hat{\beta}$ which clearly demonstrates the boundedness of the adaptation. In Figure 8, the reference trajectory is changed to demonstrates a superior a step response performance when compared with PD controller. While the step error is approaching zero for the system under the proposed adaptive controller, a steady state error is persistent with the PD controller.

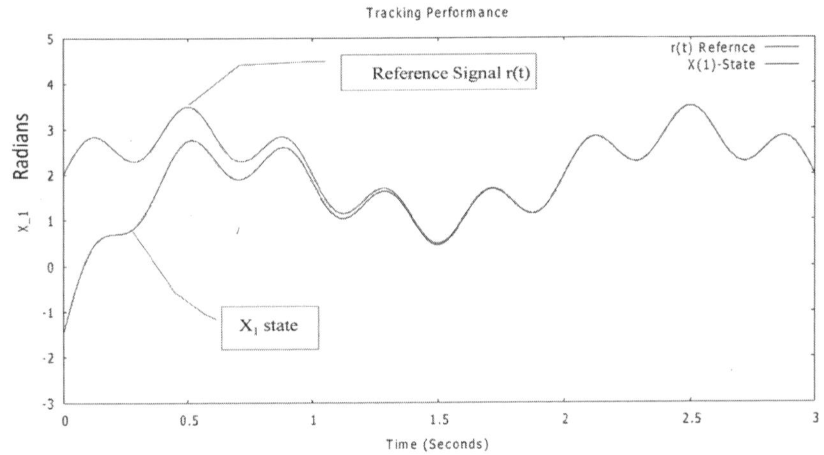

Figure 2. The tracking performance of x_1 state of the system under the proposed robust controller.

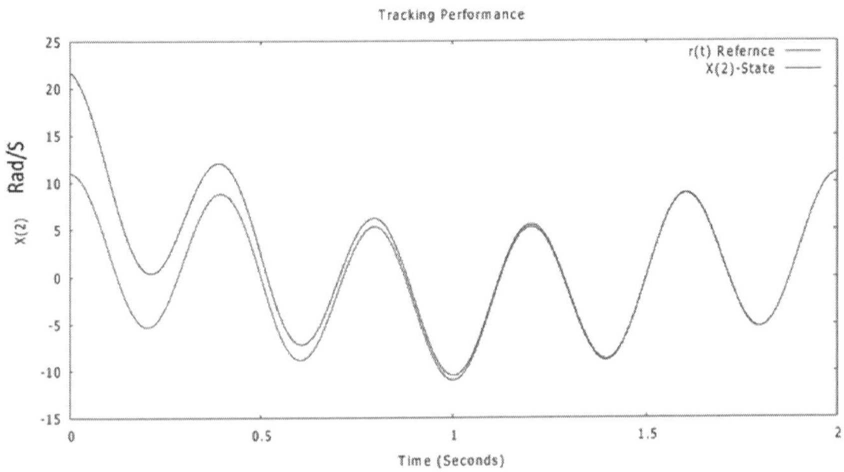

Figure 3. The tracking performance of x_2 state of the system.

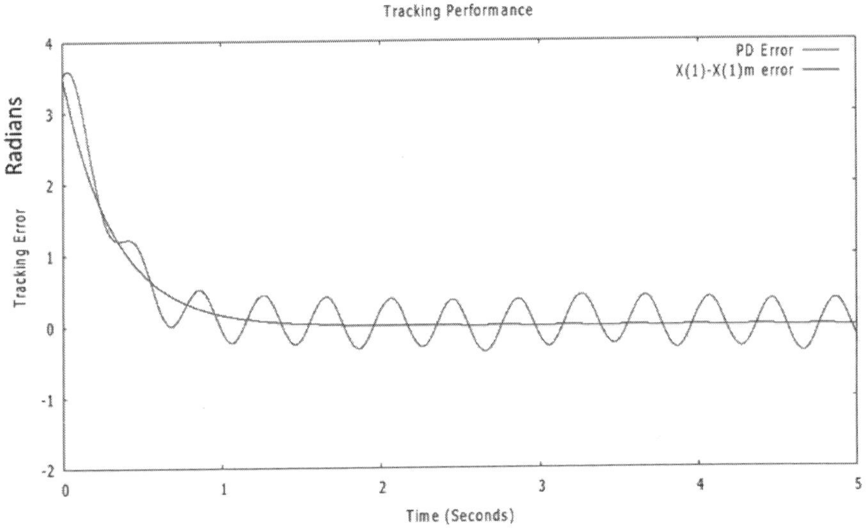

Figure 4. The adaptively compensated tracking error $\tilde{x}_1$ (blue) vs. the same tracking error of the system under a PD controller (red).

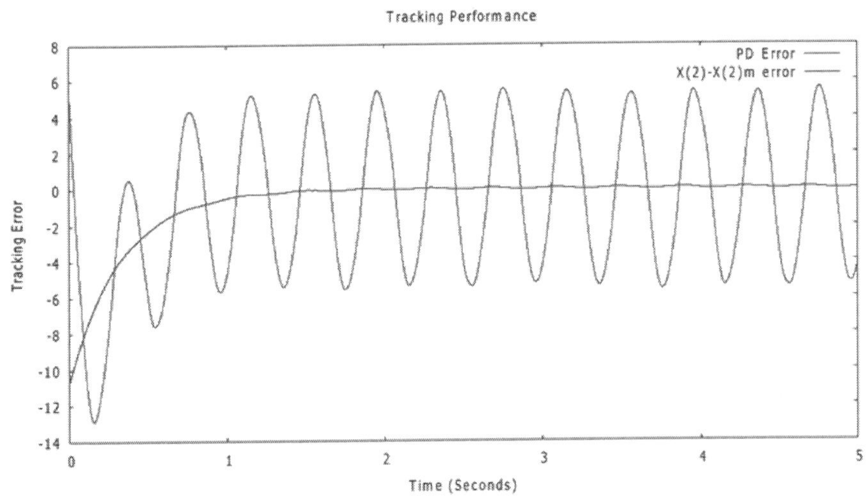

Figure 5. The adaptively compensated tracking error $\tilde{x}_2$ (blue) vs. the same tracking error of the system under a PD controller (red).

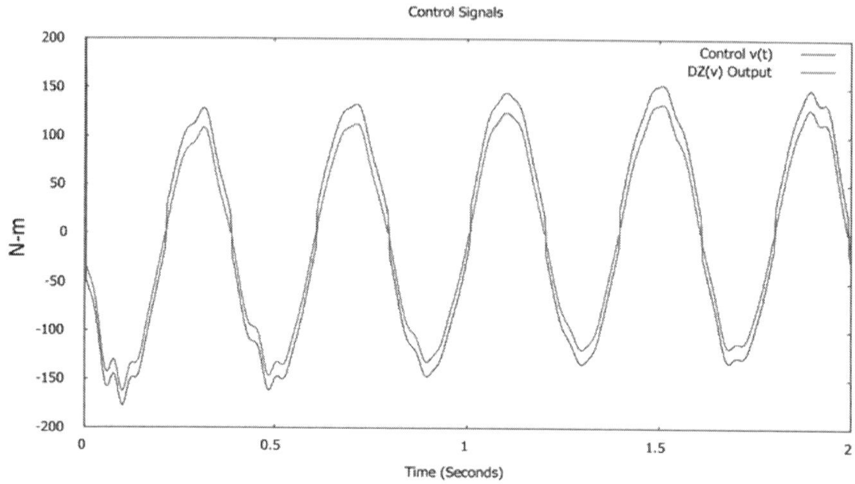

Figure 6. The control effort $DZ(u_d)$ in red vs. $v = u_d$ for the deadzone compensated system.

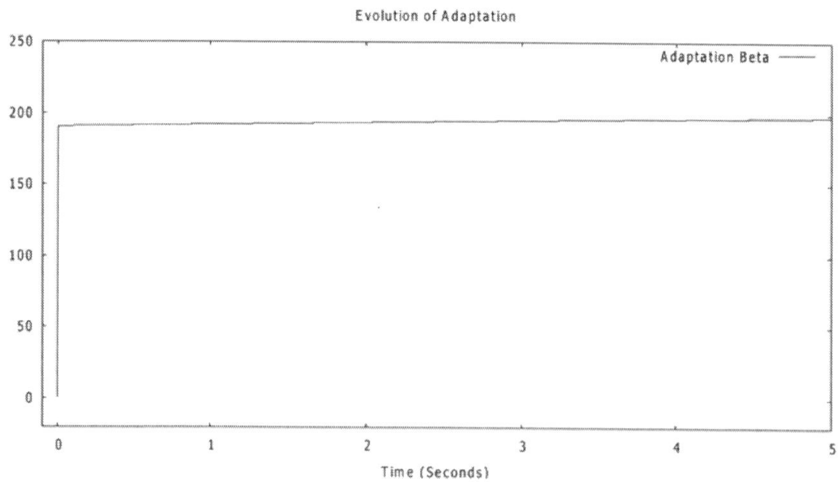

Figure 7. Evolution of the adaptation $\hat{\beta}$.

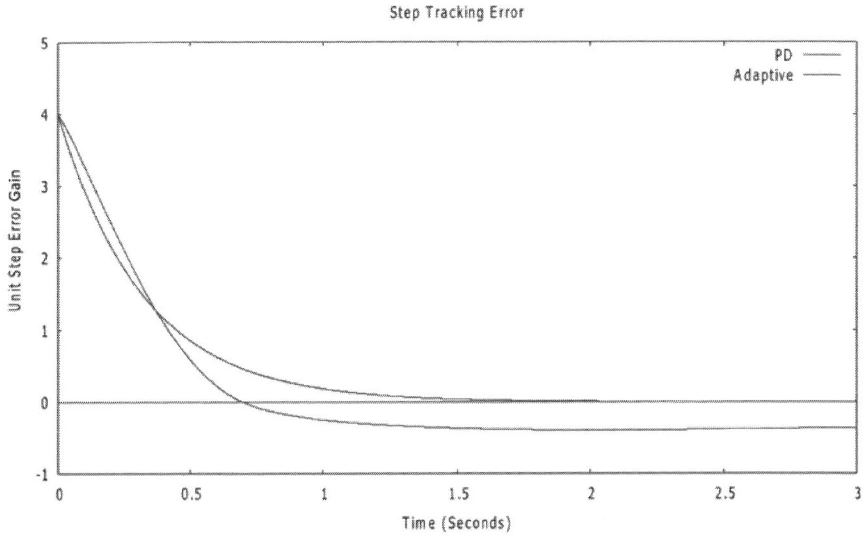

Figure 8. Step tracking performance error for $\tilde{x}_1$ for input $r(t)=1$.

5. CONCLUSION

A robust adaptive controller is used to control a class of nonlinear systems with input deadzone nonlinearity at the input. The robust controller was shown to be superior in performance when compared to a more conventional control method such as a PD controller. The system under the proposed scheme has been shown to not only ef- fectively stabilize a second order complex nonlinear system with disturbance, but also achieve asymptotic tracking. The advantage of the proposed method lies in the fact that no knowledge of the deadzone parameters needed and only an upper bound for the deadzone spacing is required. The adaptive deadzone inverse controller is smoothly differentiable and can easily be combined with any of the advanced control methodologies. The asymptotic stability of the closed-loop system has been proven by using Lyapunov arguments and simulation results confirmed the efficacy of the control methodology.

REFERENCES

1 Recker, D.A. and Kokotovic, P.V. (1993) Indirect Adaptive Nonlinear Control of Discrete-Time Systems Containing a Deadzone. Proceedings of the 32nd Conference on Decision and Control, 15-17 December 1993, San Antonio, 2647, 2653.

2 Toa, G. and Kokotovic, P. (1994) Adaptive Control of Plants with Unknown Dead-Zones. IEEE Transactions on Automatic Control, 39, 59-68.

3 Tao, G. and Kokotovic, P. (1995) Discrete-Time Adaptive Control of Systems with Unknown Dead-Zones. International Journal of Control, 61, 1-17.

4 Lewis, F.L., Tim, W.K., Wang, L.-Z. and Li, Z.X. (1999) Deadzone Compensation in Motion Control System Using Adaptive Fuzzy Logic Control. IEEE Transactions on Control Systems Technology, 7, 731-742.

5 Selmic, R.R. and Lewis, F.L. (2000) Deadzone Compensation in Motion Control Systems Using Neural Networks. IEEE Transaction of Automatic Control, 45, 602-613.

6 Wang, X.-S., Su, C.-Y. and Hong, H. (2004) Robust Adaptive Control of a Class of Nonlinear Systems with Unknown Dead-Zone. Automatica, 40, 407-413.

7 Zhonghua, W., Bo, Y., Lin, C. and Shusheng, Z. (2006) Robust Adaptive Deadzone Compensation of DC Servo System. IEE Proc.-Control Theory Application, 153, 709-713.

8 Sedighi, H.M., Shirazi, K.H. and Zare, J. (2012) Novel Equivelent Function for Deadzone Nonlinearity: Applied to Analytical Solution of Beam Vibration Using He's Parameter Expanding Method. Latin American Journal of Solids and Structures, 9, 1-10.

9 Ahmad, N.J., Alnaser, M.J. and Alsharhan, W.E. (2013) Asymptotic Tracking of Systems with Non-Symmetrical Input Deadzone Nonlinearity. International Journal of Automation and Power Engineering, 2, 287-292.

10 Ahmad, N.J., Ebraheem, H., Alnaser, M. and Alostath, J. (2011) Adaptive Control of a DC Motor with Uncertain Deadzone Nonlinearity at the Input. Control and Decision Conference (CCDC), 4295-4299.

11 Sedighi, H.M., Shirazi, K.H. and Zare, J. (2012) Novel Equivelent Function for Deadzone Nonlinearit: Appled to Analytical Solution of Beam Vibration Using He's Parameter Expanding Method. Latin American Journal of Solids and Structures, 9, 443-451.

12 Jain, S. and Khorrami, F. (1995) Robust Adaptive Control of a Class of Nonlinear Systems: State and Output Feedback. Proceeding of the 1995 American Control Conference, Seatle, 1580-1584.

13 Sankaranarayanan, S., Melkote, H. and Khorrami, F. (1999) Adaptive Variable Structure Control of a Class of Nonli- near Systems with Nonvanishing Perturbations via Backstepping. Proceedings of the American Control Conference, San Diego, 4491-4495.

14 Zhang, T.P. and Feng, C.B. (1997) Adaptive Fuzzy Sliding Mode Control for a Class of Nonlinear Systems. ACTA Automatica Sinica, 23, 361-369.

CHAPTER 8

H_∞ Based Control for Load Mitigation in Wind Turbines

Asier Diaz de Corcuera *, Aron Pujana-Arrese, Jose M. Ezquerra, Edurne Segurola and Joseba Landaluze

IKERLAN-IK4, Arizmendiarreta, 2, E-20500 Arrasate-Mondragon, The Basque Country, Spain

ABSTRACT

This article demonstrates a strategy to design multivariable and multi-objective controllers based on the H_∞norm reduction applied to a wind turbine. The wind turbine model has been developed in the GH Bladed software and it is based on a 5 MW wind turbine defined in the Upwind European project. The designed control strategy works in the above rated power production zone and performs generator speed control and load reduction on the drive train and tower. In order to do this, two robust H_∞ MISO (Multi-Input Single-Output) controllers have been developed. These controllers generate collective pitch angle and generator torque set-point values to achieve the imposed control objectives. Linear models obtained in GH Bladed 4.0 are used, but the control design methodology can be used with linear models obtained from any other modelling package. Controllers are designed by setting out a mixed sensitivity problem, where some notch filters are also included in the controller dynamics. The obtained H_∞ controllers have been validated in GH Bladed and an exhaustive analysis has been carried out to calculate fatigue load reduction on wind turbine components, as well as to analyze load mitigation in some extreme cases. The analysis compares the proposed control strategy based on H_∞ controllers to a baseline control strategy designed using the classical control methods implemented on the present wind turbines.

KEYWORDS

wind turbine; robust control; multivariable; H_∞ control; load mitigation

1. INTRODUCTION

The continuous increase of the size of wind turbines, due to the demand of higher power production installations, has led to new challenges in the design of the turbines. Moreover, new control strategies are being developed. Today's strategies trend towards being multivariable and multi-objective in order to fulfill the numerous control design specifications. To be more precise, one important specification is to mitigate loads in the turbine components to increase their life time. This can be done through the components mechanical design, the introduction of new materials or by improving the control itself. In addition to this, the behaviour of a wind turbine is non-linear, which implies that the designed control performance has to be robust.

Over the last few years, several modern control techniques used to replace the classical PI controllers (see Section 3) have been developed. These techniques are fuzzy controllers [1], adaptive control strategies [2], linear quadratic controllers [3] like the Disturbance Accommodating Control (DAC) [4] developed by NREL and tested in the CART real wind turbine [5], QFT controllers [6], Linear Parameter Varying (LPV) controllers [7] and controllers based on the H_∞ norm reduction. H_∞controllers have the capacity for robustness and these controllers are multivariable and multi-objective, so their applications in wind turbine control offer a lot of advantages and they achieve interesting results. One article dealing with this topic [8] shows the design of two controllers based on the H_∞ norm reduction applied to a simple and analytical model of a wind turbine. The first one reduces the loads on the tower with the tower fore-aft acceleration displacement measurement and controls the generator speed reference with a pitch collective control in the above rated zone. The second controller also reduces the loads on the blades with a cyclic pitch controller based on the H_∞ norm reduction. Control strategies using SISO and MISO state-space controllers based on the H_∞ norm are tested and compared in the CART3 experimental wind turbine [9]. In this article, torque controllers are used to damp the drive train mode and the tower side-to-side bending mode.

This article presents the design of two H_∞ MISO (Multi-Input Single-Output) controllers in the above rated zone (seeSection 5). These controllers not only control the generator speed and reduce the fore-aft displacements on the tower using a collective pitch controller, but they also reduce the side-to-side displacements on the tower and the loads on the drive train if a generator torque H_∞ controller is used. Furthermore, in terms of the controller design, instead of using a simple analytical model of a wind turbine, complex linear plants extracted from GH Bladed 4.0 are used, although the design methodology could be applied by using linear models obtained from any modeling package. Regarding the H_∞ controller design, some notch filters are included in the controller dynamics by means of the correct definition of the weight functions in the augmented plant of the mixed sensitivity problem. In the design process of this control strategy based on H_∞controllers, two software packages are used: GH Bladed 4.0 and MATLAB. GH Bladed is a software package commercialized by Garrad Hassan Company, commonly used by major wind turbine manufacturers to model and simulate wind turbines. The controller synthesis and discretization is carried out in MATLAB and, finally, simulations of the closed loop system are carried out using GH Bladed with different perturbed production winds. Results using H_∞ controllers are compared to the baseline controller results, based on classical control strategies, in order to do a load mitigation analysis to test the load mitigation capacity of the new designed control strategy. In the load analysis, both fatigue damage cases (DLC1.2 in IEC61400-1 Second Edition) and some extreme load cases (DLC1.6 in IEC61400-1 Second Edition) are taken into account.

2. WIND TURBINE MODEL

2.1. Non-Linear Model

The Upwind wind turbine defined in the Upwind European project was developed in GH Bladed 4.0 and it is the non-linear model used in this research project. The Upwind model consists of a 5 MW offshore wind turbine [10,11] with a monopile structure in the foundation. It has three blades and each blade has an individual pitch actuator. The rotor diameter is 126 m, the hub height is 90 m, it has a gear box ratio of 97, the rated wind speed is 11.3 m/s, the cut-out wind speed is 25 m/s and the rated rotor rotational speed is 12.1 rpm.

2.2. Linear Models

The wind turbine linear models are obtained in different operational points from the GH Bladed (version 4.0) non-linear model using the linearization tool of this software. Twelve operational points are defined from 3 m/s to 25 m/s. The Campbell diagram shows the frequencies of the structural modes of linear model family with respect to the operational points (see Figure 1).

In Table 1, the frequencies of these modes are more accurately shown for the operational point of 11 m/s wind speed and some abbreviations are defined as well as referring to the modes used throughout this article. Linear models (1) are expressed by the state-space matrices and have different inputs and outputs. Inputs are the collective pitch angle and generator torque control signals $u(t)$ and the disturbance output $w(t)$ caused by the wind speed. The outputs $y(t)$ are the sensorized measurements used to design the controller. In this case, these outputs are the generator speed w_g, the tower top fore-aft acceleration a_{Tfa} and the tower top side-to-side acceleration a_{Tss}. Due to the non-linear model complexity, and the number of modes taken into account, the order of the linear models is 55. The linear models are not reduced because, after carrying out an analysis, the best quality of the controller syntheses are obtained using high order linear plants and reducing the higher order obtained controllers:

$$\begin{aligned} \dot{X}(t) &= A \cdot X(t) + B_{11} \cdot u(t) + B_{12} \cdot w(t) \\ y(t) &= C \cdot X(t) + D_{11} \cdot u(t) + D_{12} \cdot w(t) \end{aligned} \quad (1)$$

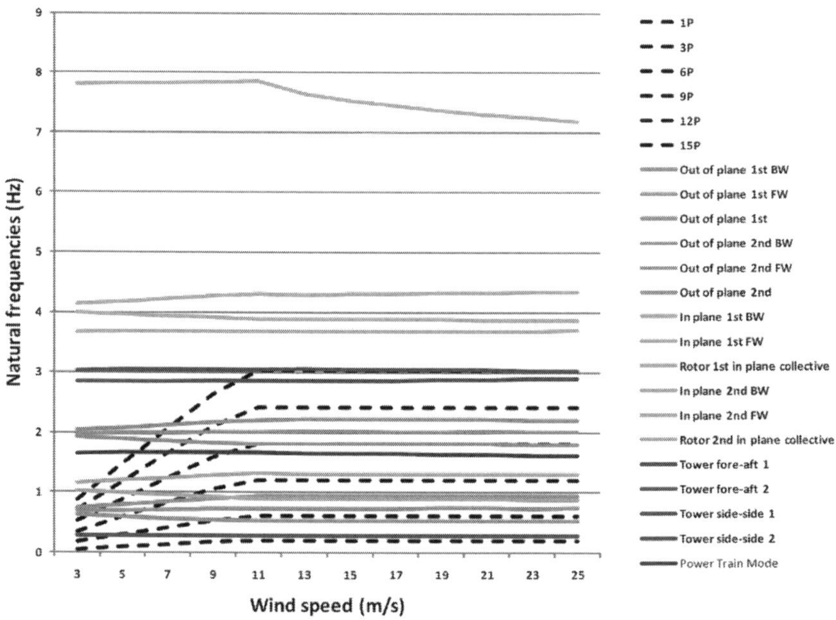

Figure 1. Campbell diagram for the Upwind model.

Table 1. Modal analysis of the Upwind model (BW: backward whirl; FW: forward whirl).

Elem.	Mode	Freq. (Hz)	Abbrev.
Rotor	In plane 1st	3.68	M_{R1ip}
	In plane 1st FW	1.31	M_{R1ipfw}
	In plane 1st BW	0.89	M_{R1ipbw}
	In plane 2st	7.85	M_{R2ip}
	In plane 2nd FW	4.30	M_{R2ipfw}
	In plane 2nd BW	3.88	M_{R2ipbw}
	Out of Plane 1st FW	0.93	M_{R1opfw}
	Out of Plane 1st	0.73	M_{R1op}
	Out of Plane 1st BW	0.52	M_{R1opbw}
	Out of Plane 2nd FW	2.20	M_{R2opfw}

Elem.	Mode	Freq. (Hz)	Abbrev.
Rotor	Out of Plane 2nd	2.00	M_{R2op}
	Out of Plane 2st BW	1.80	M_{R2opbw}
Drive Train	Drive Train	1.66	M_{DT}
Tower	1st tower side-to-side	0.28	M_{T1ss}
	1st tower fore-aft	0.28	M_{T1fa}
	2nd tower side-to-side	2.85	M_{T2ss}
	2nd tower fore-aft	3.05	M_{T2fa}
Non-str.	1P	0.2	$1P$
	3P	0.6	$3P$

3. BASELINE CLASSICAL CONTROL STRATEGY (C1)

The wind turbine control strategy is defined by a curve (see Figure 2) which relates the generator torque and the generator speed [12]. Three control zones are distinguished in this curve: below rated zone, transition zone and above rated zone. In the below rated zone the control objective is to maintain the power coefficient (C_p) in the optimum value. In the Upwind baseline controller, this is done by means of a generator torque control depending on the generator speed measurement (2). The generator torque T_{br} is proportional to the square of the generator speed by a constant K_{opt}.

$$T_{br} = K_{opt} \cdot w_g^2$$
$$K_{opt} = 2.14\left[\frac{\text{Nm}}{(\text{rad/s})^2}\right] \qquad (2)$$

The aim in the transition zone is the control of generator speed by varying the generator torque. In the Upwind model, this can be done with a torque proportional-integral *PI* (3) controller [13] or with an open loop torque control which produces a ramp [14] to relate the generator torque and the generator speed. In the C1 control strategy, the *PI* values in the transition zone (wind speed of 11 m/s) used in the Upwind baseline controller are K_{pt} and K_{it} (3), where $u(s)$ is the generator torque control signal and $e(s)$ is the generator speed error:

$$u(s) = \frac{K_{pt}s + K_{it}}{s} \cdot e(s) \qquad (3)$$
$$K_{pt} = 4200; K_{it} = 2100$$

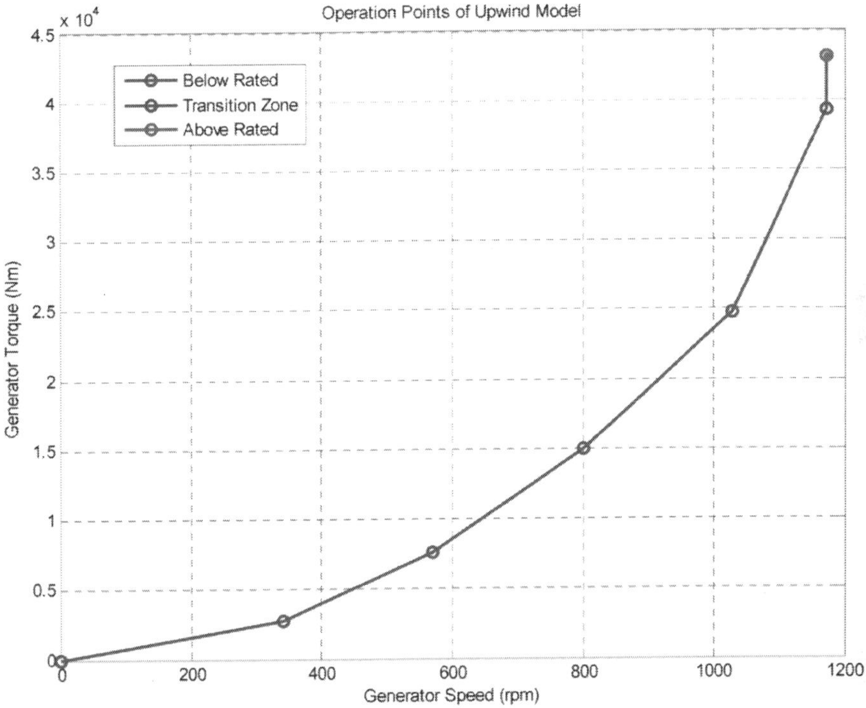

Figure 2. Curve of power production control zones for the Upwind wind turbine.

In the above rated zone, the goal is the generator speed control at the nominal value of 1173 rpm varying the collective pitch angle in the blades to maintain the electric power at the value of 5 MW. To do this, a gain-scheduled (*GS*) *PI* controller [15] is used. In this case, the controller input $u(s)$ is the generator speed error, and the controller output $\beta_{col}(s)$ is the collective pitch angle control signal. The linear plants used to tune the gain-scheduled *PI* controller are the plants which relate pitch angle and generator speed. These plants have different gains, so gain-scheduling is used to guarantee the stability of the closed loop system in spite of the gain differences. To develop the gain-scheduling, two *PI* controllers (4) in two operational points, winds of 13 m/s and 21 m/s, are tuned:

$$
\begin{aligned}
K_{pt\_13} &= 0.009; K_{it\_13} = 0.003 \\
K_{pt\_21} &= 0.0039; K_{it\_21} = 0.0013
\end{aligned}
\tag{4}
$$

In the other operational points, the *PI* parameters are extrapolated by a first order approximation. A similar gain-scheduling strategy is proposed in [14]. Instead of using the wind speed signal from the anemometer, this *PI* is scheduled by the collective pitch angle in the blades. The corresponding steady-state collective pitch angle is 6.42° for the operational point with a wind speed of 13 m/s, and the corresponding steady-state collective pitch angle value is 18.53° for the operational point with a wind speed of 21 m/s. Finally, some series notch filters are useful to improve the *PI* controller response [16]. Some design criteria are established to tune these controllers in these operational points:

1 Output sensitivity peak: 6 dB approximately.

2 Open loop phase margin between 30 and 60 degrees.

3 Open loop gain margin between 6 and 12 dB.

4 To maintain constant the PI zero frequency.

The drive train damping filter (DTD) is included. The aim of the DTD is to reduce the wind effect on the drive train mode [15,17]. The DTD for the Upwind model (5) consists of one gain, with one differentiator, one real zero and a pair of complex poles:

$$T_{DTD}(s) = \left[K_1 \cdot \frac{s\left(1+\frac{1}{w_1}s\right)}{\left(\left(\frac{1}{w_2}\right)^2 s^2 + 2\xi_2 \frac{1}{w_2} s + 1\right)} \right] \cdot w_g(s) \qquad (5)$$

$K_1 = 641.45$ Nms/rad; $w_1 = 193$ rad/s; $w_2 = 10.4$ rad/s; $\xi_2 = 0.984$

The input of the filter is the generator speed w_g and the output is a contribution T_{DTD} to the generator torque set-point signal. Finally, the tower fore-aft damping filter (TD) is designed to reduce the wind effect on the tower first fore-aft mode in the above rated power production zone [15,17]. For the Upwind baseline controller, the filter

(6) consists of a gain with one integrator, a pair of complex poles and a pair of complex zeros:

$$B_{fa}(s) = K_{TD} \cdot \frac{1}{s} \cdot \left[\frac{1+(2 \cdot \varsigma_{T1} \cdot s / w_{T1}) + (s^2 / w_{T1}{}^2)}{1+(2 \cdot \varsigma_{T2} \cdot s / w_{T2}) + (s^2 / w_{T2}{}^2)} \right] \cdot a_{Tfa}(s) \tag{6}$$

$K_{TD} = 0.035;\ w_{T1} = 1.25$ rad/s; $\varsigma_{T1} = 0.69;\ w_{T2} = 3.14$ rad/s; $\varsigma_{T1} = 1$

The input of the filter is the fore-aft acceleration measured in the tower top a_{Tfa} and the output is a pitch contribution $\beta_{fa}$to the collective pitch angle. In conclusion, the baseline control strategy is defined in Figure 3. Other strategies to reduce the loads on the wind turbine can be developed, but they are not included in the considered baseline controller.

4. OBJECTIVES FOR DESIGNING THE NEW PROPOSED CONTROL STRATEGY

The control objectives for the developed wind turbine control strategy working in the above rated power production zone are as follows:

1 Generator speed control (increase of the output sensitivity bandwidth and reduction of the peak in comparison with the baseline controller).

2 To mitigate the load on the drive train reducing the wind effect on the drive train mode.

3 To mitigate the load on the tower reducing the wind effect on the tower first modes (side-to-side and fore-aft).

4 To improve the load mitigation in comparison to a baseline controller based on the classical baseline control strategy.

To achieve these control objectives, a generator speed sensor and an accelerometer on the tower top are used [18].

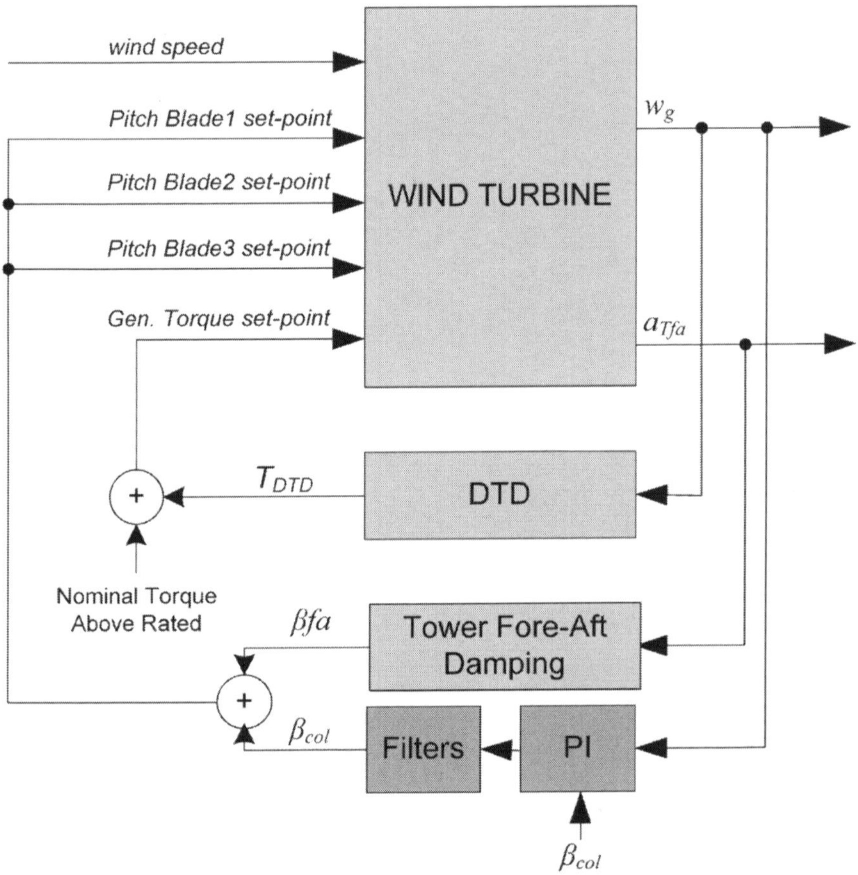

Figure 3. Baseline C1 control strategy.

5. NEW PROPOSED CONTROL STRATEGY BASED ON H$_\infty$ NORM REDUCTION (C2)

5.1. Design of Control Strategy Based on H$_\infty$ Norm Reduction

This strategy consists of two robust, multivariable and multi-objective controllers based on the H$_\infty$ norm reduction (seeFigure 4). The generator torque controller and the pitch controller are designed separately [19]. The torque controller has two inputs (generator speed w_g and tower top side-to-side acceleration a_{Tss}) and one output (generator torque control signal$T_{H\infty}$). On the other hand, the pitch controller has two inputs (generator speed w_g and tower top fore-aft acceleration a_{Tfa}) and one output (collective pith control signal $\beta_{H\infty}$). The collective pitch angle set-point value is the pitch control signal $\beta_{H\infty}$. However, the generator torque set-point value is the addition of the generator torque control signal $T_{H\infty}$ and the generator torque nominal value in the above rated zone.

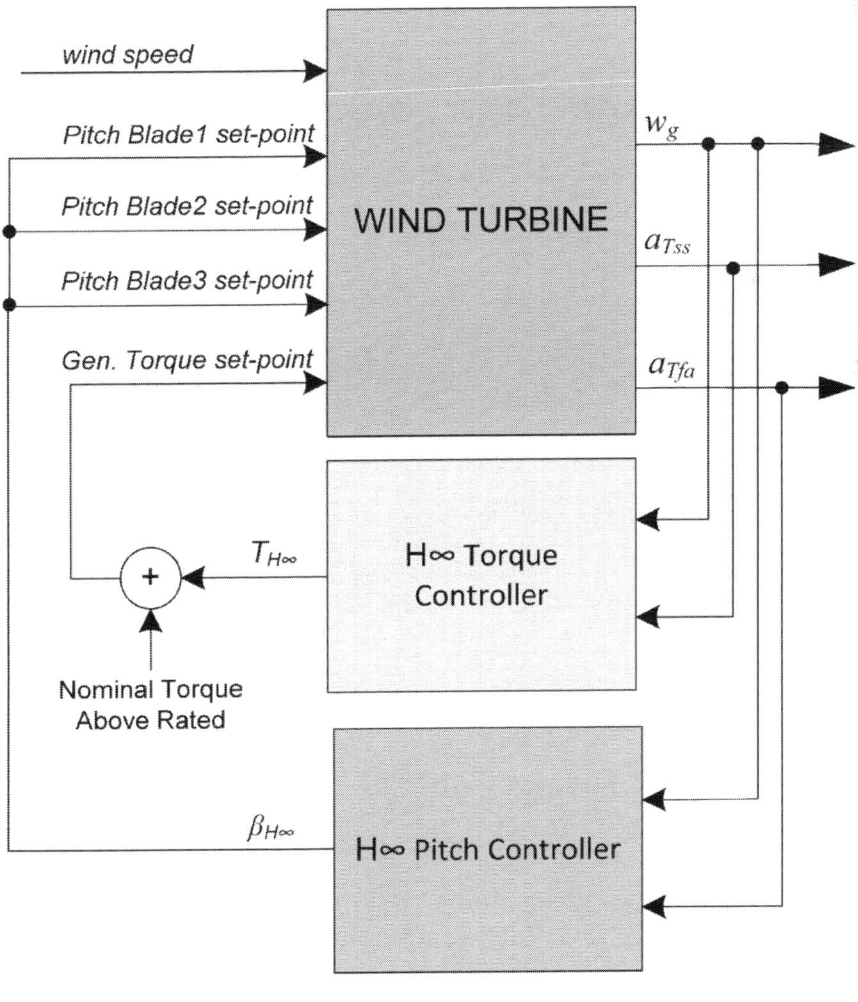

Figure 4. C2 control strategy based on the H_∞ norm reduction.

The control design method can be divided into the following steps:

1. To extract the wind turbine linear models from the GH Bladed non-linear model. The wind turbine used for this design is the 5 MW Upwind model.
2. To analyze the linear models in Simulink extracting the Campbell Diagram.
3. To design the torque H_∞ controller in MATLAB.
4. To design the pitch H_∞ controller in MATLAB taking into account the previous designed H_∞ torque controller.
5. To analyze the controller robustness in MATLAB.
6. To test the controllers in Simulink.
7. To include the controllers in the GH Bladed External Controller.
8. To simulate the GH Bladed non-linear model using the designed two MISO H_∞ controllers.
9. To compare the time domain and frequency domain results to the baseline classical controller.
10. To analyze the fatigue loads and extreme loads reduction of the proposed control strategy compared to the baseline control strategy.

5.2. Generator Torque Controller (H_∞ Torque Controller)

1. The designed generator torque controller based on the H_∞ norm reduction solves two of the control objectives proposed in Section 4:
2. To reduce the wind effect on the drive train mode M_{DT}.
3. To reduce the wind effect on the tower side-to-side mode M_{T1ss}.

To design the controller, a mixed sensitivity problem (7) will be solved. The nominal plant *G*(*s*) is selected at the operational point of 19 m/s wind speed and has one input *T* (generator torque), two outputs w_g and a_{Tss} and 55 states (seeFigure 5). $G_{11}(s)$ is the plant with a generator torque input and a generator speed output, while $G_{12}(s)$ is the plant with a generator torque input and a tower top side-to-side acceleration output. p_1 and p_2 are the disturbance outputs of the plant, *u*is the control signal, y_1 and y_2 are the controller inputs, and Zp_{11}, Zp_{12}, Zp_2, Zp_{31} and Zp_{32} are the performance outputs. The augmented plant (see Figure 6) of this mixed sensitivity problem is scaled using the constants Du, De_1, De_2, Dp_1 and Dp_2 (8).

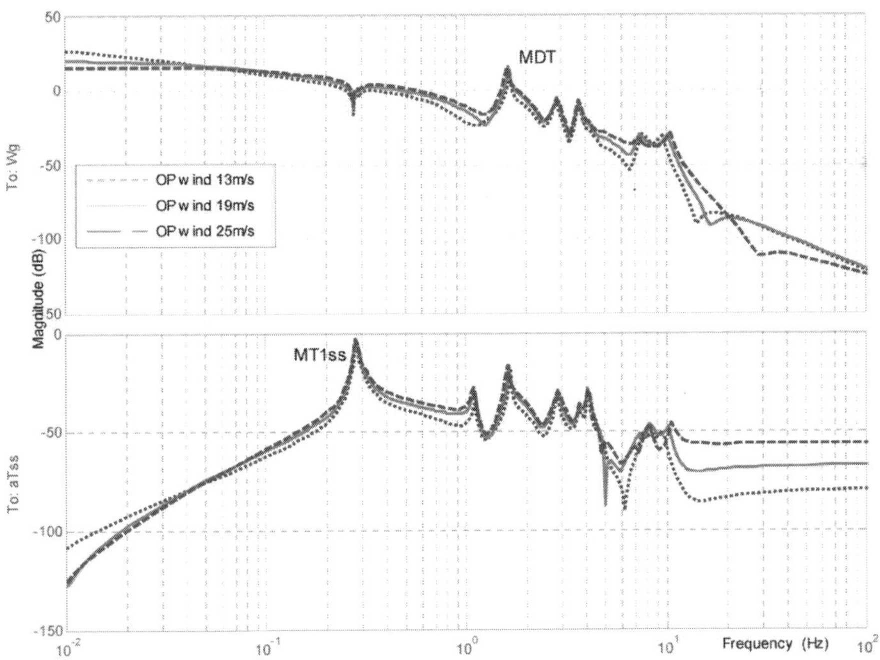

Figure 5. Family of plants for the H∞ torque controller design.

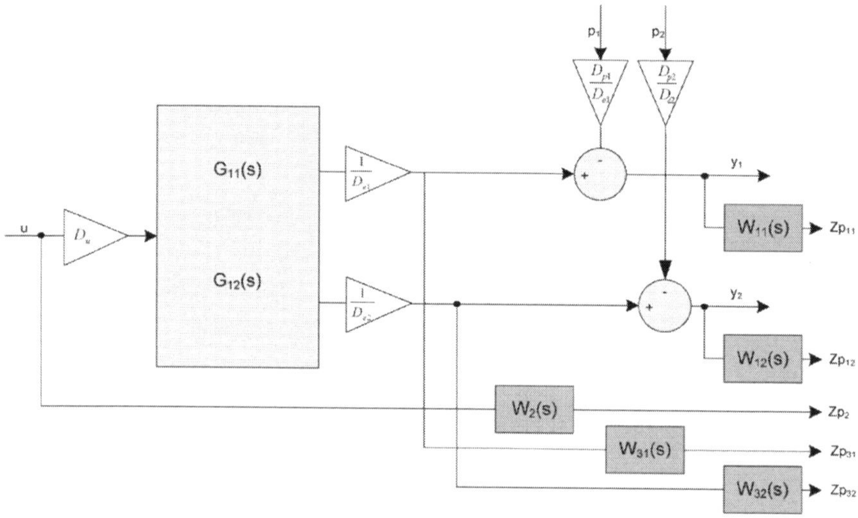

Figure 6. Augmented plant for the MISO Mixed-Sensitivity Problem.

$$\begin{pmatrix} Zp_{11} \\ Zp_{12} \\ Zp_2 \\ Zp_{31} \\ Zp_{32} \\ y_1 \\ y_2 \end{pmatrix} = \begin{pmatrix} -\frac{Dp_1}{De_1}\cdot W_{11} & 0 & \frac{Du}{De_1}\cdot G_{11}(s)\cdot W_{11} \\ 0 & -\frac{Dp_2}{De_2}\cdot W_{12} & \frac{Du}{De_2}\cdot G_{12}(s)\cdot W_{12} \\ 0 & 0 & W_2 \\ 0 & 0 & \frac{Du}{De_2}\cdot G_{11}(s)\cdot W_{31} \\ 0 & 0 & \frac{Du}{De_1}\cdot G_{12}(s)\cdot W_{32} \\ -\frac{Dp_1}{De_1} & 0 & \frac{Du}{De_1}\cdot G_{11}(s) \\ 0 & -\frac{Dp_2}{De_2} & \frac{Du}{De_2}\cdot G_{12}(s) \end{pmatrix} \cdot \begin{pmatrix} p_1 \\ p_2 \\ u \end{pmatrix} \tag{7}$$

The uncertainties of the family of plants are not considered in this mixed sensitivity problem due to the fact that the drive train and tower modes frequencies do not considerably vary in the above rated zone. The weight functions (9) W_{31} and $W_{32}$are not used, so their values are 1 in order not to consider them in the MATLAB Robust Toolbox [20]. W_{11} is an inverted notch filter centred on the M_{DT} frequency and W_{12} is another inverted notch filter centred on

(C2)

the M_{T1ss} frequency. W_2 is an inverted low-pass filter used to reduce the controller activity in high frequencies (see Figure 7):

$$Du = 90; De_1 = 0.1; De_2 = 1; Dp_1 = 0.1; Dp_2 = 1 \tag{8}$$

$$W_{11}(s) = \frac{(s^2 + 6.435s + 104.9)}{(s^2 + 0.1416s + 104.9)}$$

$$W_{12}(s) = \frac{(s^2 + 9.984s + 3.117)}{(s^2 + 0.04437s + 3.117)} \tag{9}$$

$$W_2(s) = \frac{30000(s + 5.027)}{(s + 6.823e5)}$$

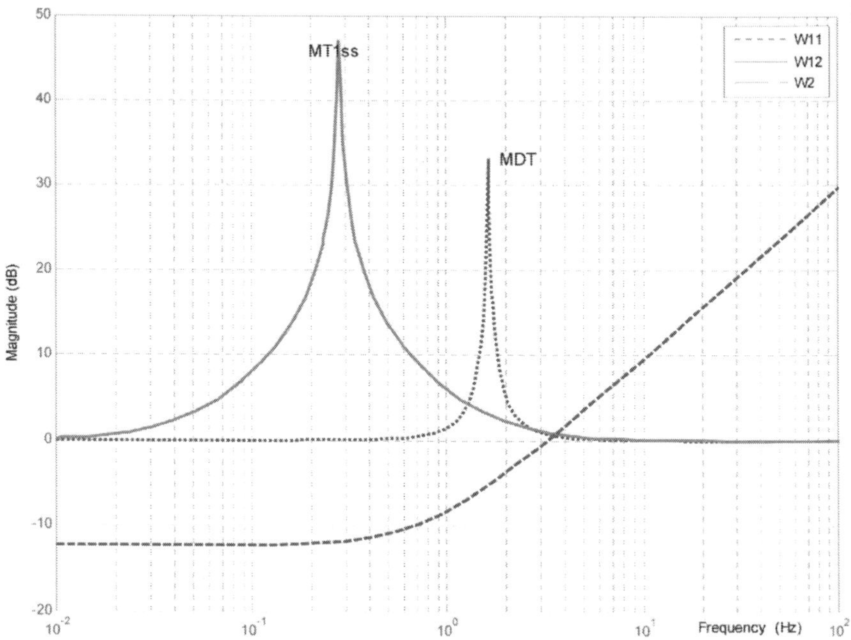

Figure 7. Weight functions for H∞ torque controller design.

After doing the controller synthesis, the obtained controller (see Figure 8) has to be re-scaled to adapt the inputs and the outputs to the real non-scaled plant. The obtained controller order is 39 but, finally, the controller order is reduced to order 25 and discretized using a sample time of 0.01 s. The discretized controller is represented by the state space matrices A_{TD},B_{TD}, C_{TD} and D_{TD} (10):

$$X_{TD}(k+1) = A_{TD} \cdot X_{TD}(k) + B_{TD} \cdot \begin{pmatrix} e_{wg}(k) \\ a_{Tss}(k) \end{pmatrix}$$
$$T_{H\infty}(k) = C_{TD} \cdot X_{TD}(k) + D_{TD} \cdot \begin{pmatrix} e_{wg}(k) \\ a_{Tss}(k) \end{pmatrix} \tag{10}$$

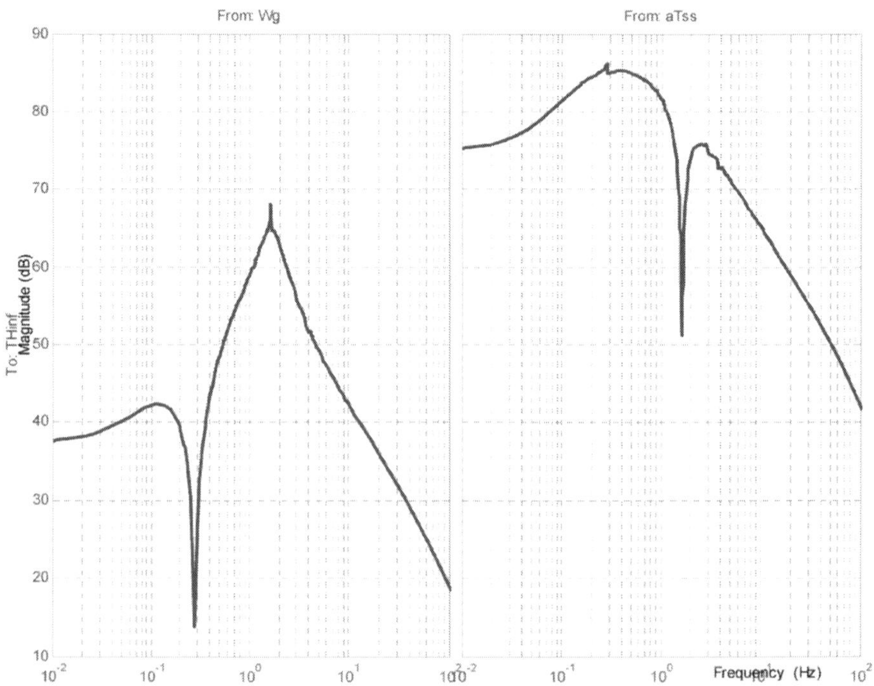

Figure 8. H∞ Torque Controller.

5.3. Collective Pitch Angle Controller (H_∞ Pitch Controller)

The collective pitch H_∞ MISO controller solves the other control objectives:

1 Generator speed control increasing the closed loop disturbance attenuation bandwidth.

2 Reduction of the wind effect on the tower fore-aft mode.

3 Inclusion of notch filters at particular frequencies in the controller dynamics to mitigate other excited frequencies (seeTable 2) in the nominal plant.

Table 2. Frequency of notch filters in the H_∞ Pitch Controller.

Mode	Freq. (*Hz*)
$1P$	0.20
$3P$	0.60
M_{T2ss}	2.86
M_{R1ip}	3.69
M_{R2ip}	7.36

Another mixed sensitivity problem is proposed to develop this controller. In this case, the nominal plant $GI(s)$ is selected for the operational point of 19 m/s wind speed (see Figure 9), it has one input β (collective pitch angle), two outputs w_g anda_{Tfa} and considers the coupling caused by the inclusion of the H_∞ MISO torque controller designed in the previous section.$G_{11}(s)$ is the plant with a collective pitch input and a generator speed output and $G_{12}(s)$ is the plant with a collective pitch input and the tower top fore-aft acceleration output. This control scenario has new scaled constants (11) and the family of plants is considered as an additive uncertainty model due to the variations of the linear plants according to the operational point in the above rated zone:

$$Du = 1; De_1 = 10; De_2 = 0.1; Dp_1 = 10; Dp_2 = 0.1 \tag{11}$$

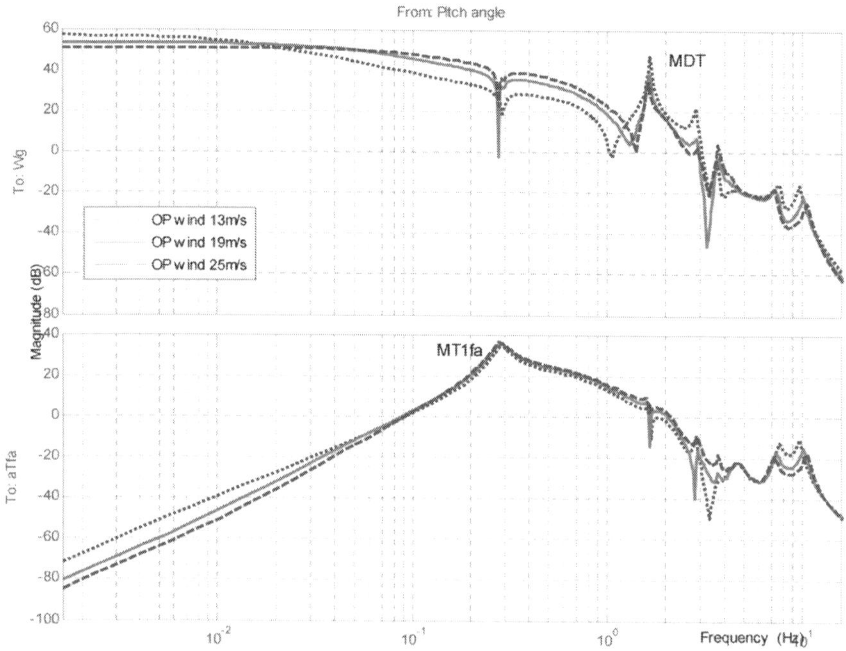

Figure 9. Family of plants for the H∞ pitch controller design.

Regarding the weight functions (12) in this mixed sensitivity problem, the W_{11} is an inverted high-pass filter which determines the desired profile of the output sensitivity function. W_{12} is an inverted notch filter centred on the M_{T1fa} and $W_2$is an inverted low-pass filter used to reduce the controller activity in high frequencies, including some inverted notch filters centred on excited frequencies (see Table 2) to include notch filters in the pitch controller dynamics (see Figure 10).

$$W_{11}(s) = \frac{(s+125.7)}{(s+6.283e-5)}$$

$$W_{12}(s) = \frac{(5s^4+5.733s^3+31.58s^2+18s+49.28)}{(s^4+0.3117s^3+6.288s^2+0.9786s+9.856)} \tag{12}$$

$$W_2(s) = \frac{200000(s+628.3)(s^2+0.1005s+1.579)(s^2+0.3016s+14.21)}{(s+6.283e5)(s^2+0.02011s+1.579)(s^2+0.06032s+14.21)} \cdot \frac{(s^2+1.438s+322.9)(s^2+1.885s+537.5)(s^2+3.7s+2139)}{(s^2+0.2875s+322.9)(s^2+0.371s+537.5)(s^2+0.7399s+2139)}$$

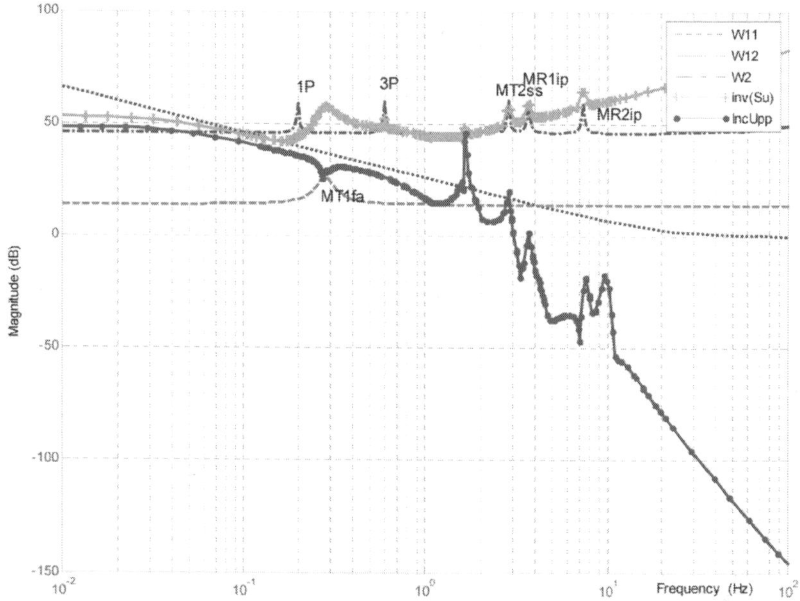

Figure 10. Weight functions for the H∞ pitch controller design.

The gains of the upper uncertainty model IncUpp are bounded by W_2 weight functions (see Figure 8) to guarantee the robust controller design. After re-scaling the obtained controller (see Figure 11), whose order is 45, it is reduced to order of 24 and discretized using a sample time of 0.01 s. The discretized controller is represented by the state space matrices A_{BD},B_{BD}, C_{BD} and D_{BD} (13):

$$
\begin{aligned}
X_{BD}(k+1) &= A_{BD} \cdot X_{BD}(k) + B_{BD} \cdot \begin{pmatrix} e_{wg}(k) \\ a_{Tfa}(k) \end{pmatrix} \\
\beta_{H\infty}(k) &= C_{BD} \cdot X_{BD}(k) + D_{BD} \cdot \begin{pmatrix} e_{wg}(k) \\ a_{Tfa}(k) \end{pmatrix}
\end{aligned} \qquad (13)
$$

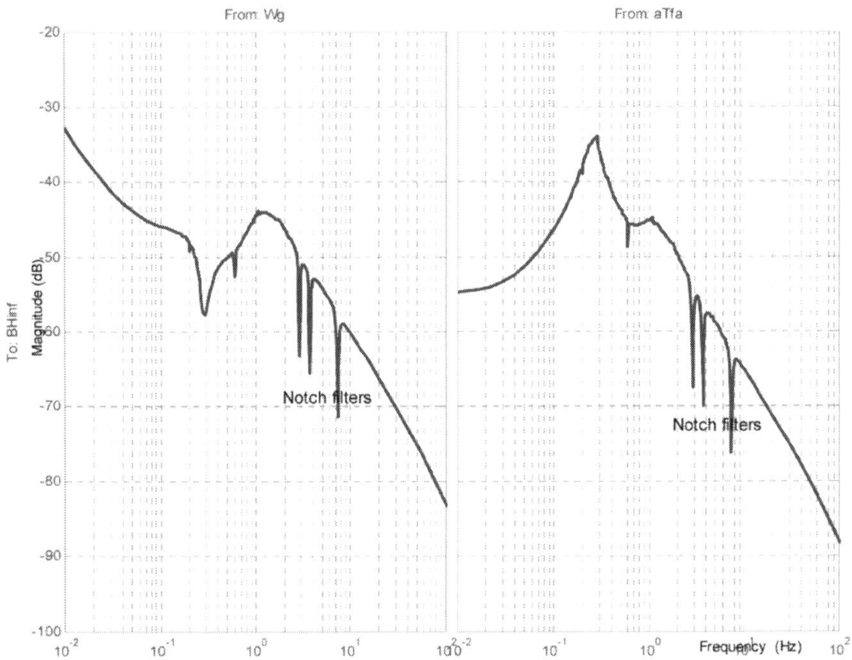

Figure 11. H$_\infty$ Pitch Controller.

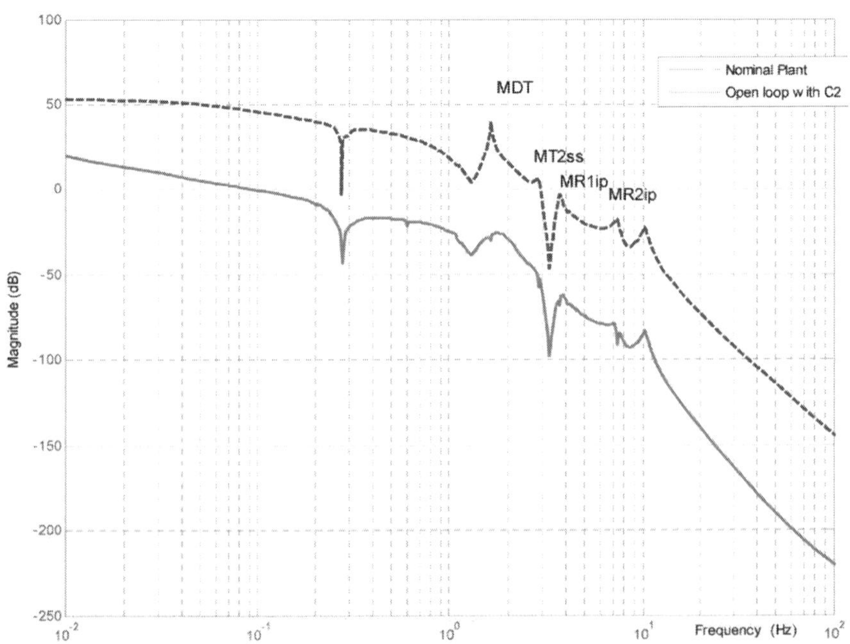

Figure 12. Effect of the pitch controller notch filter in open loop.

5.4. Analysis of the H_∞ Control Strategy

Gain variations in the generator speed control are only considered in the controller robust analysis due to the fact that the tower and drive train modes for the Upwind model have constant frequencies in the above rated zone. The controller robustness is guaranteed because the gains of the upper uncertainty model IncUpp are bounded by the inverse of the control sensitivity function Su [21] (see Figure 10). To compare the response of the designed controllers to the baseline controller, the two control strategies in the above rated zone are considered:

- **C1**: Baseline control strategy with drive train damping filter and tower fore-aft damping filter activated (see Figure 3).
- **C2**: Proposed control strategy with two H_∞ MISO controllers.

Using the C2 control strategy, the generator speed output disturbance attenuation bandwidth DABW in the different operational points is higher than using the baseline control strategy (see Table 3) and the generator speed output disturbance attenuation peak *DAP* is lower near the designed wind speed nominal operational point of 19 m/s. For the nominal plant, for which the controller is designed, the generator speed output sensitivity function (see Figure 13) shows the peak and the bandwidth to control the generator speed output for a generator speed output disturbance. This sensitivity function clearly shows the increase in the bandwidth achieved with the H_∞ C2 control strategy. These improvements are of interest in order to reduce extreme loads, as will be shown in section 6.The reduction of the wind effect on the M_{DT} drive train mode is critical for the control strategy design, so it has been designed first. This mode reduction appears in the wind effect on different parts of the wind turbine due to the hard coupling of this mode in the system. For example, Figure 12shows the M_{DT} mitigation in the plant which relates the generator speed frequency response for a pitch angle input. Figure 14 shows the frequency response of the tower side-side acceleration for a wind input where the drive train mode is mitigated clearly with the C1 and C2 control strategies compared to the plant without the control system.

Table 3. Comparison of generator speed disturbance attenuation.

OP (*m/s*)	C1		C2	
	DABW (*Hz*)	DAP (*dB*)	DABW (*Hz*)	DAP (*dB*)
13	0.037	6.06	0.035	3.35
15	0.045	6.06	0.044	3.59
17	0.052	6.09	0.057	4.31
19	0.058	6.31	0.070	5.29
21	0.061	6.00	0.078	5.78
23	0.065	6.05	0.089	6.70
25	0.069	6.04	0.10	7.84

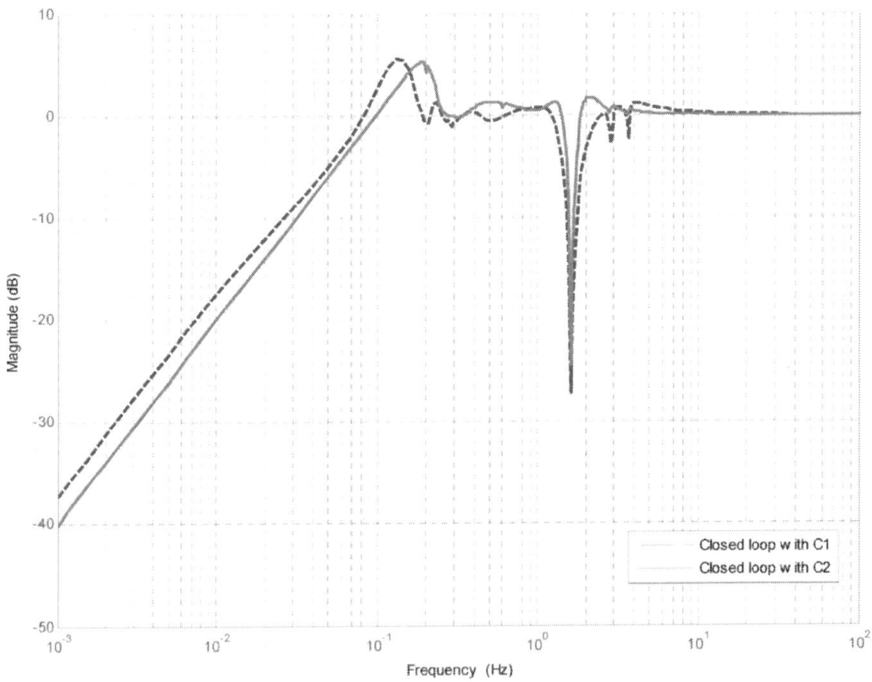

Figure 13. Generator speed output sensitivity function.

To analyze the reduction of the fore-aft and side-to-side tower accelerations due to the mitigation of the wind effect on these modes using the H_∞ control strategy, the closed loop response has to be analyzed in time and frequency domains for the two control strategies. The frequency response of the tower top side-to-side acceleration for a wind input (see Figure 14) is mitigated at the M_{T1ss} frequency using the

C2 control strategy, but this mode is not reduced when the C1 strategy is used because it was not designed for that. This gain reduction at this frequency involves an amplitude reduction of the tower side-to-side acceleration in time domain (see Figure 14). The frequency response of the tower top fore-aft acceleration for a wind input (see Figure 15) is mitigated at the M_{T1fa} frequency using the C1 and C2 control strategies. This mode is not very excited in the Upwind model, but this mitigation could be more useful in other wind turbine models. This gain reduction at the peak of the M_{T1fa} mode involves an amplitude reduction of the tower fore-aft acceleration in time domain. The gain mitigation on the M_{T1fa} and M_{T1ss} frequencies is important to reduce the momentums in the tower y and x axis respectively, and the momentum reductions mean load mitigations on the tower.

Finally, the notch filters included in the pitch controller dynamics are used to mitigate other excited frequencies which appear in the plant which relates the pitch angle with the generator speed. In this plant (see Figure 12), for the operational point of wind 19 m/s, the structural modes M_{T2ss}, M_{R1ip}, M_{R2ip} are excited and they are mitigated using the H_∞ C2 control strategy. The effect of the notch filters at the frequencies $1P$ and $3P$ (blade passing frequencies) are only observed in the frequency analysis of the time domain simulations, because the excitations in these frequencies are not expressed in the linear plants extracted from GH Bladed.

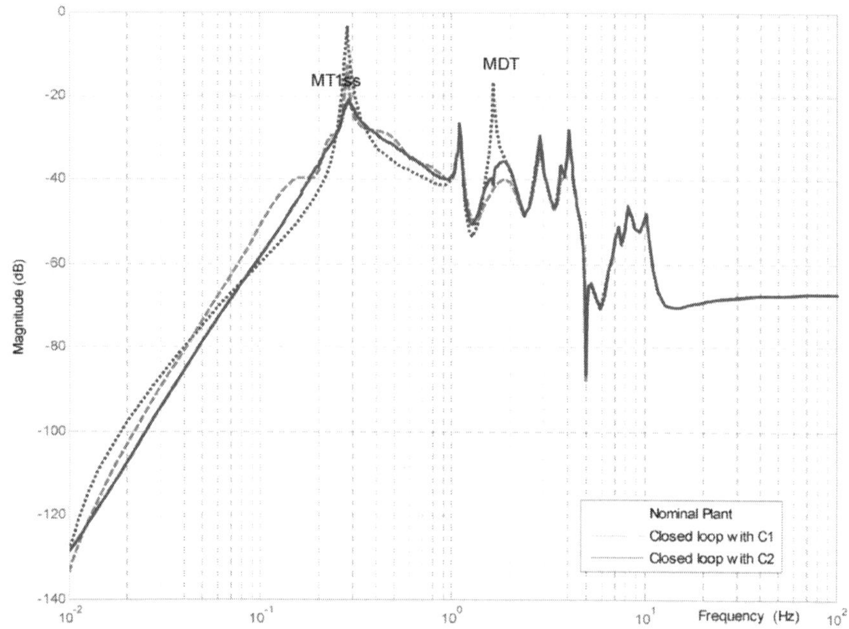

(a)

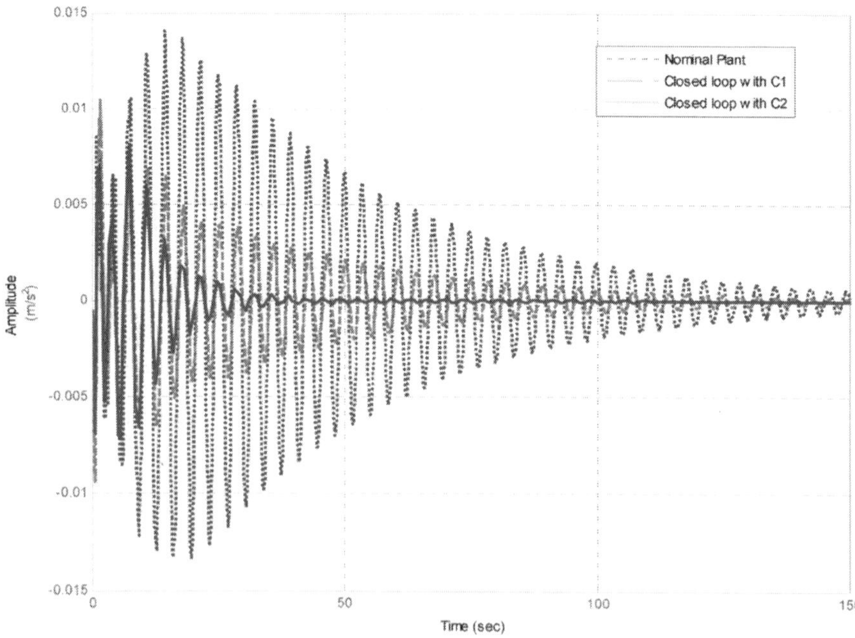

(b)

Figure 14. Tower top side-to-side acceleration response for a wind input (**a**) Frequency response; (**b**) Time domain wind step response.

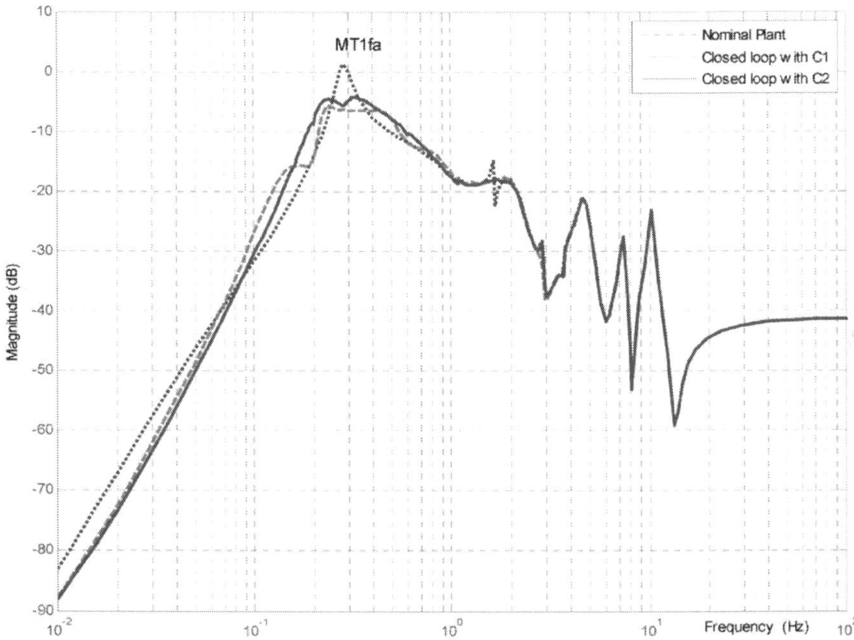

(a)

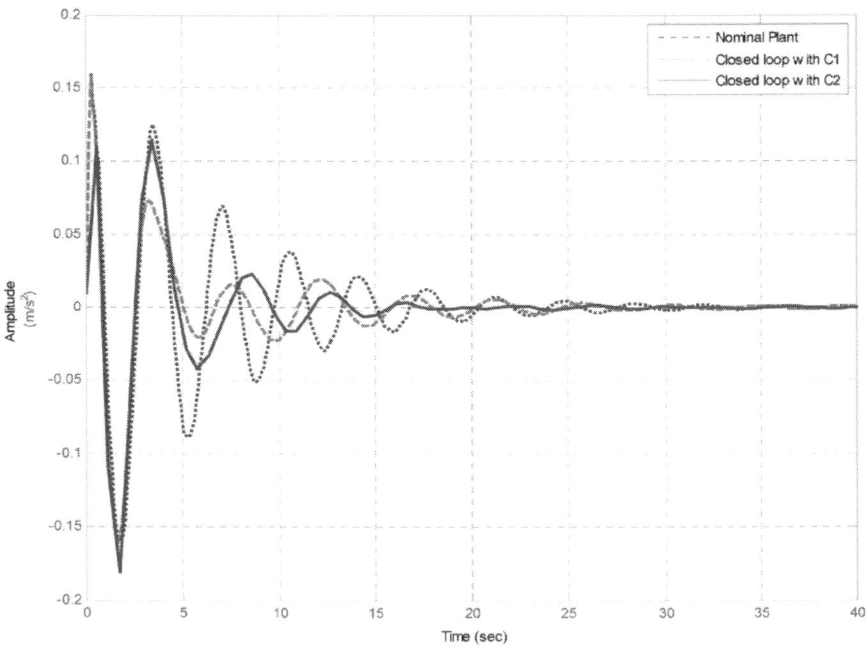

(b)

Figure 15. Tower top fore-aft acceleration response for a wind input (**a**) Frequency response; (**b**) Time domain wind step response.

6. RESULTS IN GH BLADED

6.1. External Controller in GH Bladed

The two H_∞ MISO controllers are included in the External Controller in GH Bladed to do time domain simulations using the Upwind wind turbine non-linear model. The External Controller [22] is the name of the programmed code to control the wind turbine non-linear model in GH Bladed. GH Bladed calls to the External dynamic library .dll with the frequency determined by the sample time of the control strategy. The C1 and C2 control strategies are included in the External Controller to carry out the control strategies in the above rated power production zone. However, the control strategy in the below rated and transition zones is the same as the baseline control strategy and it is described in section 3. For a more realistic comparison between the results using the C1 control strategy and the results using the C2 control strategy in the above rated zone, the C2 strategy is divided into two cases:

- **C2.1**: The accelerometer to measure the tower top side-to-side acceleration is disabled. This is done to compare C1 and C2 control strategies without tower side-to-side damping.
- **C2.2**: The tower top side-to-side accelerometer is activated and C2 control strategy works without sensor signals restrictions (see Figure 4).

In the control strategy based on the H_∞ MISO discretized controllers, the control signals are calculated for each sample time (0.01 s) using the present vector of states expressed in the state-representation of the controller dynamics. The sample time has been selected after consulting wind turbine manufacturer references. The strategy to calculate the controller output is divided into four steps:

1 To initialize the controller state-space matrices A, B, C, and D from a static library and initialize the actual state vector$X(k)$.

2 To update the present vector of controller inputs $e(k)$ reading the wind turbine measurements from the sensors.

3 To calculate the vector of present controller outputs $u(k)$ using matrices C, D and the current vectors of controller inputs $e(k)$ and states $X(k)$.

4 To calculate the vector of the next sample time controller states $X(k)$ using matrices A and B and the actual vectors of controller inputs $e(k)$and states $X(k)$. In the next sample time this vector of controller states will be the current vector of controller states.

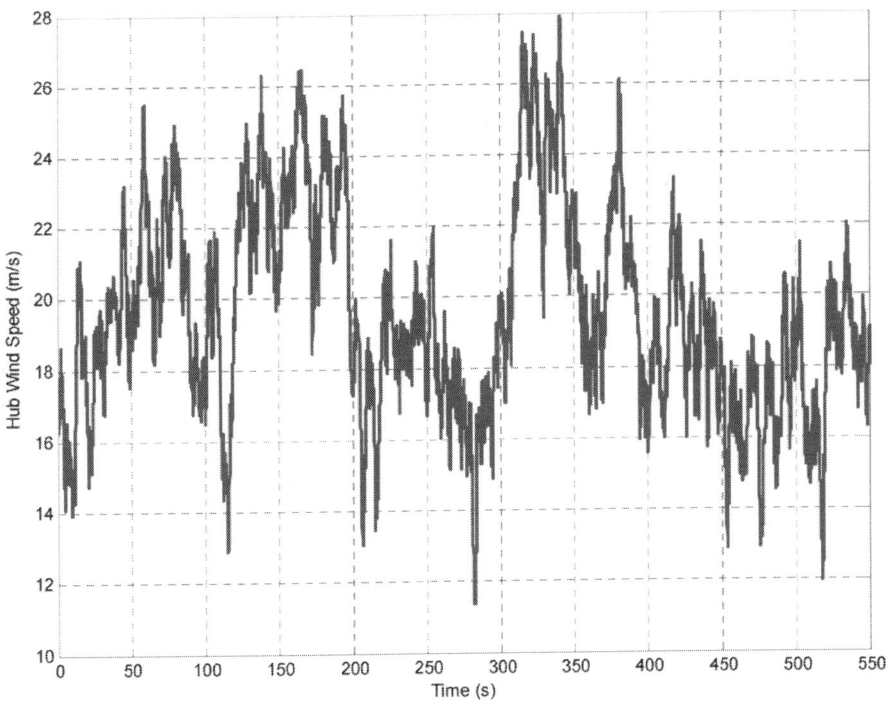

Figure 16. Hub wind speed for a turbulent production wind of 19 m/s.

6.2. Fatigue Analysis (DLC1.2 in IEC61400-1 Second Edition)

The rain flow counting algorithm [23,24] is used to analyze the load reduction capacity of the designed controllers. A fatigue analysis is carried out using this algorithm to determine the fatigue damage on the wind turbine components. The fatigue damage analysis, called load equivalent analysis, follows these steps:

1 To carry out time domain simulations using the non-linear wind turbine model and the designed controller. Twelve simulations of 600 s have been carried out using odd production winds from mean speeds from 3 m/s to 25 m/s.

2 To subject some signals of loads in time simulations (stationary hub Mx, stationary hub My, tower base Mx, tower base My, blade MFlap and blade MEdge) to the rain flow counting algorithm (one for each measured variable) using the toolbox in MATLAB [25] to carry out this analysis.

3 To obtain the load equivalent L_{eq} (14) for each kind of material and for each simulated wind. The material is defined by the m value. m is the slope of the SN curve of the material, where S is the fatigue strength and N the number of cycles to failure. N_i, the number of cycles, and L_i, the cycles amplitudes, are extracted from the rainflow counting and N_{rd} is the number of points of the time domain simulation. For glass fibre $m = 10$, for cast modular iron $m = 7$ and for welded steel$m = 3$:

$$L_{eq} = \left(\frac{\sum (n_i \cdot L_i)}{N_{rd}} \right)^{\frac{1}{m}} \qquad (14)$$

1. The twelve simulations must be taken into account to calculate the total load equivalent for each material. The load equivalent referring to the Weibull distribution w_{eqm} (15) is calculated for each wind and each material. The total load equivalent for one material L_{eqw} (16) referring to the Weibull distribution is calculated with the summation of the w_{eqm}.w_c is a parameter of the Weibull distribution, *slife* is the standard life of wind turbines (20 years) and *tsim* is the simulated time of the considered variable in this load equivalent analysis:

$$w_{eqm} = L_{eq}{}^{m} \cdot w_c \cdot slife \,/\, tsim \qquad (15)$$

$$L_{eq} = \left(\sum w_{eqm}\right)^{(1/m)} \qquad (16)$$

2. To compare the wind turbine life variations $comp_{life}$ (17) between two compared load equivalent analysis. L_{eqw1} is the total load equivalent value for twelve simulations and L_{eqw2} is the other total load equivalent value for the other twelve simulations:

$$comp_{life} = \frac{slife}{\left(\dfrac{L_{eqw1} - L_{eqw2}}{100}\right)^{m}} \qquad (17)$$

In Figure 17 the controlled signals (generator speed and electric power) are compared for a turbulent production wind of 19 m/s (see Figure 16) using the C1, C2.1 and C2.2 control strategies. The generator speed is controlled for the nominal value 1173 rpm and the electrical power around 5 MW. The power spectral density (*PSD*) performs a frequency analysis of time domain simulations. In Figure 18 the control signals (collective pitch angle and generator torque) are compared. The time domain simulation shows the quick response of the pitch angle using the C2.1 and C2.2 strategies and the torque contribution signal of the C2.2 control strategy to mitigate the side-to-side displacement on the tower. Figure 19 shows the reduction of the wind effect on the tower fore-aft mode in the tower base momentum My and the reduction of the wind effect on the tower side-to-side mode and drive train mode in tower base momentum Mx.

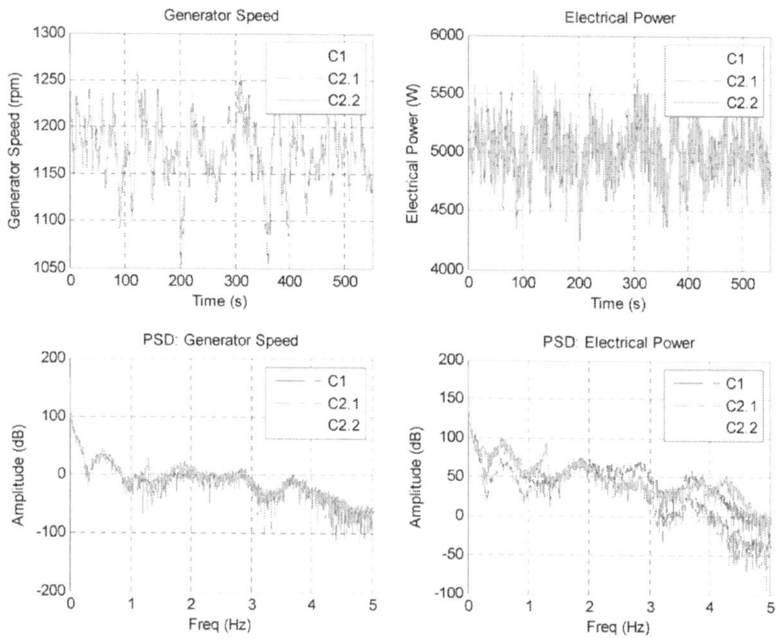

Figure 17. Generator speed and electrical power with a turbulent production wind of 19 m/s.

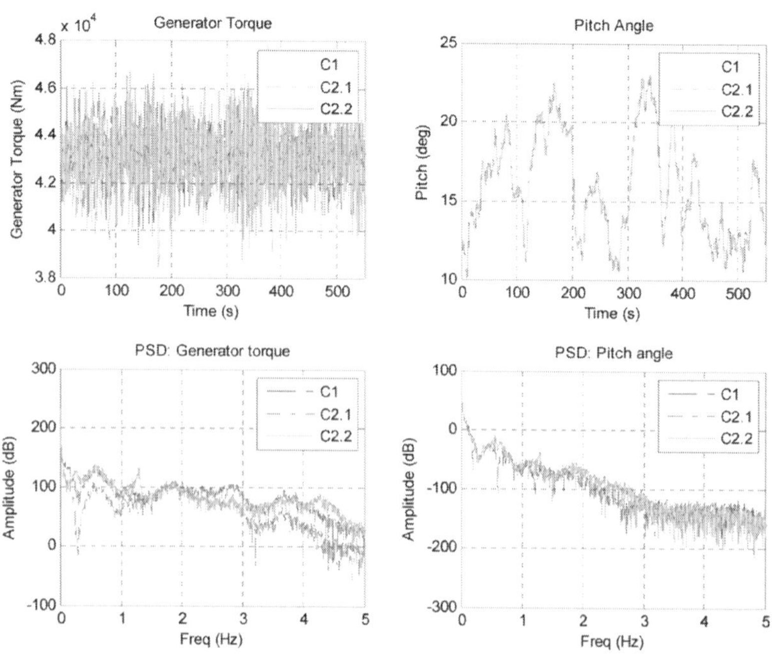

Figure 18. Generator torque and pitch angle with a turbulent production wind of 19 m/s.

Finally, the fatigue damage analysis of the components of the wind turbine is carried out as a result of the load equivalent analysis. Table 4 shows the load reduction percentage of the control strategies on the different components of the Upwind model. C1 is the reference strategy to compare the load reduction achieved with C2.1 and C2.2 control strategies. Results are calculated for three material constants, m, which are commonly used by the commercial companies of wind turbines. If $m = 3$ using the two H_∞ MISO controllers, the load reduction is 0.4% on the Stationary Hub Mx momentum using the C2.2 control strategy and 4.8% using the C2.1 strategy, 13.4% on the Tower Base Mx momentum with the C2.2 and 2.6% using the C2.1, and 4.7% on the Tower Base My momentum respect to the C1 baseline controller using the C2.2 and 5.2% with the C2.1. If $m= 9$, the load reduction is 3.1% on the Stationary Hub Mx momentum, 19.6% on the Tower Base Mx momentum and 8.8% on the Tower Base My momentum using the C2.2 control strategy. On the other hand, if $m = 9$ using the C2.1 control strategy, the load reduction is 4.2% on the Stationary Hub Mx momentum, −0.1% on the Tower Base Mx momentum and 10.9% on the Tower Base My momentum. If $m = 12$ using the C2.2 strategy, the load reduction is 2.9% on the Stationary Hub Mx momentum, 19.2% on the Tower Base Mx momentum and 10.6% on the Tower Base My momentum. For this m value using the C2.1 control strategy, the load reduction is 4.0% on the Stationary Hub Mx momentum, −0.2% on the Tower Base Mx momentum and 13.6% on the Tower Base My momentum. Load reduction figures less than 0.4% should not be considered because they can be caused by mathematical calculation precision in the load equivalent algorithm (in blue colour in Table 4).

In these simulations there is an excitation of the rotor in-plane 1st FW mode M_{R1ipfw} (1.2 Hz). This excitation is not critical from the point of view of the load reduction in a wind turbine system as is proven in this load equivalent analysis. The cause of this excitation is the bandwidth of the torque controller. The torque controller reduces the wind effect on the drive train mode M_{DT} (1.6 Hz) and tower 1st side-to-side mode M_{T1ss} (0.28 Hz). The torque H_∞ MISO controller (see Figure 8) dynamics from tower top side-to-side acceleration to torque set point value introduces a high gain in frequencies between 0.2 Hz and 1.6 Hz which produces the In plane 1st FW mode excitation. To reduce this excitation, a notch filter in the rotor in-plane 1st FW frequency must be included in the weight functions W_2 used to design the torque controller.

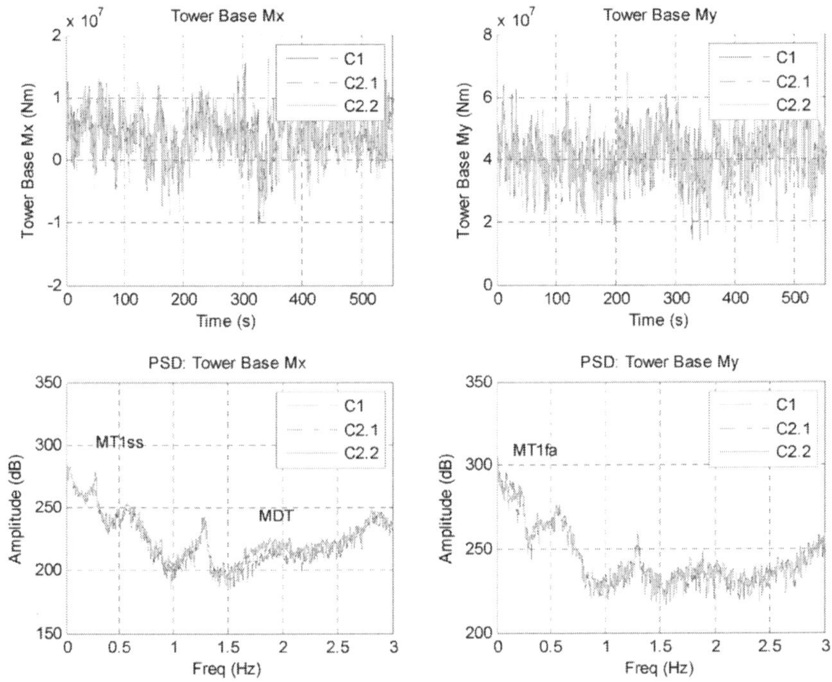

Figure 19. Loads on tower base with a turbulent production wind of 19 m/s.

6.3. Extreme Load Analysis (DLC1.6 in IEC61400-1 Second Edition)

The extreme load DLC1.6 analysis studies the system response for different kinds of extreme gusts. This analysis is divided into three different steps:

1 To carry out time domain simulations using the non-linear wind turbine model and the C1 and C2 different control strategies. Six simulations of different kinds of gusts have been carried out. The gusts are called Vr-0, Vr-p, Vr-n, Vout-0, Vout-p, Vout-n.

2 To analyze the six simulations and extract the maximum value of the generator speed signal and some momentums (tower base Mx, tower base My, tower base Mxy, hub total bending Myz, blade MFlap and blade MEdge).

3 To compare these maximum values using the C2 control strategy with regard to the baseline C1 control strategy.

Table 4. Comparison of the load equivalent analysis.

	m	C1-C2.1 (%)	C1-C2.2 (%)
	3	4.8	0.4
Stationary Hub Mx	9	4.2	3.1
	12	4.0	2.9
	3	0.2	0.2
Stationary Hub My	9	0.9	1
	12	1.2	1.5
	3	4.8	0.4
Gearbox Torque	9	4.2	3.1
	12	4.0	2.9
	3	2.6	13.4
Tower Base Mx	9	−0.1	19.6
	12	−0.2	19.2
	3	5.2	4.7
Tower Base My	9	10.9	8.8
	12	13.6	10.6
	3	0.1	−0.2
Blade1 MFlap	9	0.1	−0.1
	12	0.1	−0.2
	3	0	0.1
Blade1 MEdge	9	0	0
	12	0	0

Other extreme load cases (as for instance DLC1.5 in IEC61400-1 Second Edition cases) are not taken into account because results depend especially on the stop strategy, which has not been implemented. Moreover, a safety control strategy is not developed in the supervisory control to stop the wind turbine in generator over-speed cases. This article only shows the response of the developed C1 and C2 control strategies in the above rated zone without taking into account any safety control strategy.

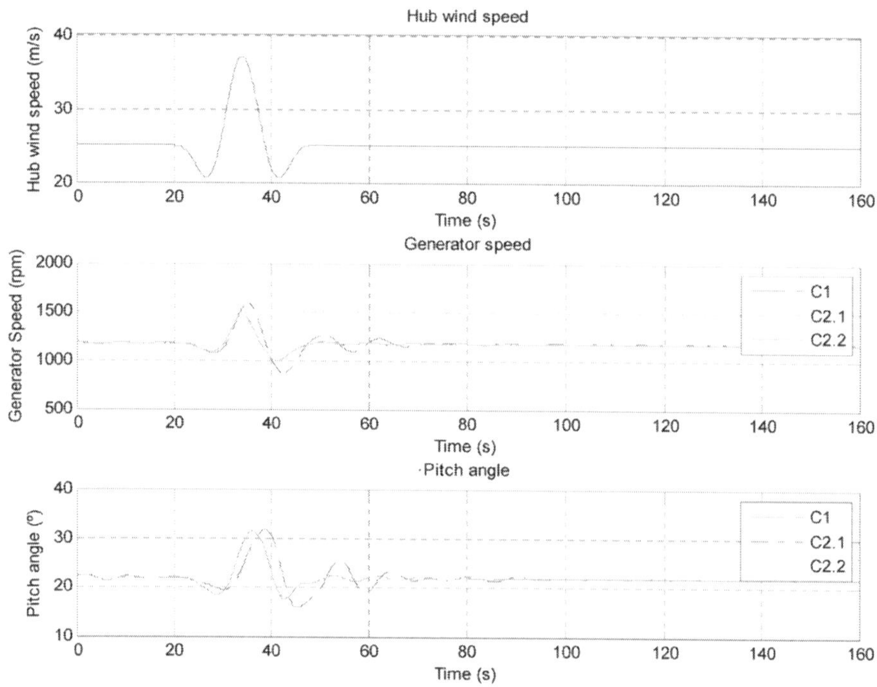

Figure 20. Vout-p extreme gust simulation.

The results of the extreme load DLC1.6 analysis are summarized in Table 5. The C2.1 and C2.2 control strategies, based on the H_∞ norm reduction, give better results than the baseline control strategy C1. Using C2.2, the reduction of the generator speed maximum value, compared to the C1 strategy, is 7.8%, the tower base Mx momentum is reduced by 7.8% and the blade MEdge momentum is reduced by 26.3%. The tower base My momentum, tower base Mxy momentum and hub total bending Myz momentum are reduced by around 1.5% when using the C2.2 control strategy. Using C2.1, the reduction of the generator speed maximum value is 7.8%, the tower base Mx momentum is not reduced and the blade MEdge momentum is reduced by 25.9%. However, the tower base My momentum, tower base Mxy momentum and hub total bending Myz momentum are not considerably reduced using the C2.1 control strategy. The blade MEdge momentum is considerably reduced due to the increase in the generator speed output disturbance bandwidth and the reduction in the peak of this sensitivity function. In Figure 20, the results of the simulation in GH Bladed using an input of wind of a Vout-p gust and using the C1 control strategy are compared to the same wind input using the C2 control strategy without including any safety system to avoid generator over-speeds. The higher bandwidth of the generator speed output sensitivity function obtained with the C2 control strategy means a quick response of the collective pitch angle signal and the consistent reduction of the generator speed maximum value. Furthermore, the mitigation of the variations of the generator speed using the H_∞ controllers is the main reason for the mitigation of the extreme load on the blade MEdge momentum. The most useful advantage of this reduction of extreme loads using the H_∞ control strategy is to avoid the activation of special safety strategies to stop the wind turbine. Usually, these special safety strategies are activated when the generator speed is higher than a critical value and involves losses of the electric power and critical momentary increases of loads in the wind turbine.

Table 5. Comparison of the extreme load analysis.

	C1	C2.1	C1-C2.1 (%)	C2.2	C1-C2.2 (%)
Generator Speed	1589 rpm	1464 rpm	7.86	1465 rpm	7.8
Tower Base Mx	29278 KNm	29285 KNm	0.02	26983 KNm	7.8
Tower Base My	158258 KNm	155500 KNm	1.74	155473 KNm	1.7
Tower Base Mxy	158311 KNm	155500 KNm	1.77	155555 KNm	1.7
Hub total bending Myz	12991 KNm	12780 KNm	1.62	12817 KNm	1.3
Blade MFlap	18341 KNm	18400 KNm	−0.32	18355 KNm	−0.07
Blade MEdge	9946 KNm	7366 KNm	25.94	7327 KNm	26.3

7. CONCLUSIONS

The work carried out and presented in this article can be summarized as follows:

1 The offshore Upwind 5 MW wind turbine model is developed using the GH Bladed 4.0 software package.

2 A classical control strategy for wind turbines is defined and it is considered the baseline controller for comparing with the new developed control strategies.

3 New design process of a control strategy based on H_∞ controllers is defined and validated in GH Bladed. The new strategies are applied in above rated power production zone in wind turbines. The results obtained in the closed loop simulations using GH Bladed 4.0 software package show the fatigue load reduction on the desired components (tower and drive train) compared to the classical baseline control strategy (see Figure 21). Using the designed H_∞ controllers, the extreme load reduction in case DLC1.6 does not appear only in the tower, but also in the three blades. Results obtained using H_∞ controllers have these outstanding benefits from the load reduction point of view due to some interesting properties:

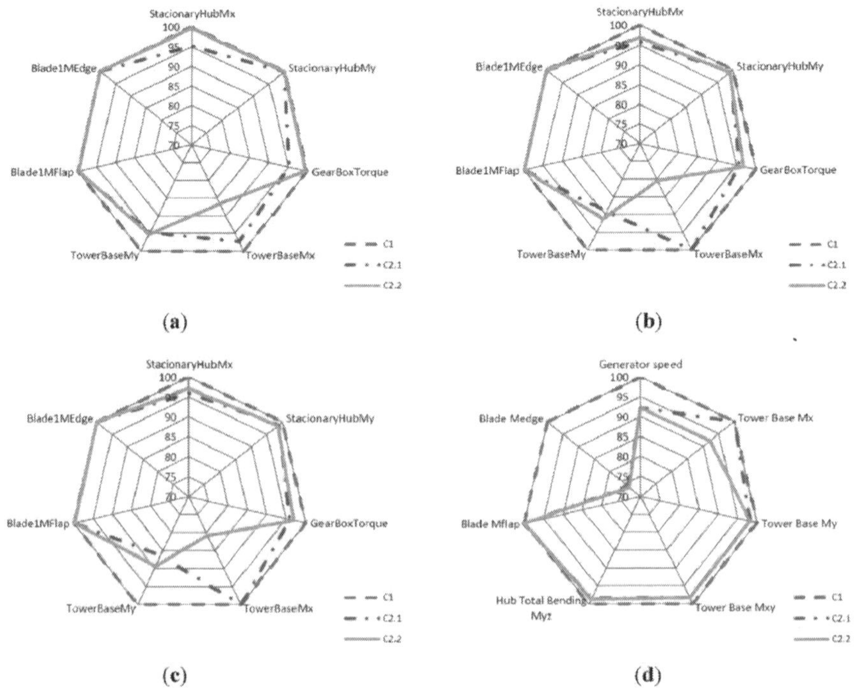

Figure 21. Review of load mitigation (**a**) Equivalent load analysis m = 3; (**b**) Equivalent load analysis m = 9; (**c**) Equivalent load analysis m = 12; (**d**) Extreme load analysis.

- The attenuation of the generator speed output disturbance bandwidth is higher than that obtained using the classical control strategy.
- The attenuation of the generator speed output disturbance peak is higher than that obtained using the classical control strategy.
- The proposed control strategy based on the H_∞ norm reduction takes into account the coupling between variables in the wind turbine system. The designed controller is multivariable and multi-objective.
- The controller robustness is guaranteed due to the small gain theorem properties applied to the H_∞ controller synthesis.
- Some notch filters can be included in the controller dynamics using a correct definition of the mixed sensitivity problem. This is very useful for reducing excited modes on non-desired frequencies.

ACKNOWLEDGEMENTS

The work described in this paper has been supported in part by the Basque Country Government (Spain) and the Regional Council of Aquitaine (France), in the frame of Cooperation Commons Funds Euskadi-Aquitaine (Project Bladed).

REFERENCES

1 Caselitz, P.; Geyler, M.; Giebhardt, J.; Panahandeh, B. Hardware-in-the-Loop development and testing of new pitch control algorithms. In Proceeding of European Wind Energy Conference and Exhibition (EWEC), Brussels, Belgium, March 2011; pp. 14–17.

2 Johnson, K.E.; Pao, L.Y.; Bạlas, M.J.; Kulkarni, V.; Fingersh, L.J. Stability analysis of an adaptive torque controller for variable

speed wind turbines. In Proceeding of IEEE Conference on Decision and Control, Atlantis, Bahamas, December 2004; pp. 14–17.

3 Nourdine, S.; Díaz de Corcuera, A.; Camblong, H.; Landaluze, J.; Vechiu, I.; Tapia, G. Control of wind turbines for frequency regulation and fatigue loads reduction. In Proceeding of 6th Dubrovnik Conference on Sustainable Development of Energy, Water and Environment Systems, Dubrovnik, Croatia, September 2011; pp. 25–29.

4 Wright, A.D. *Modern Control Design for Flexible Wind Turbines*; Technical Report for NREL: NREL/TP-500-35816; Colorado, CO, USA; July; 2004.

5 Wright, A.D.; Fingersh, L.J.; Balas, M.J. Testing state-space controls for the controls advanced research turbine. In Proceeding of 44th AIAA Aerospace Sciences Meeting and Exhibit, Reno, NV, USA, January 2006; pp. 9–12.

6 Sanz, M.G.; Torres, M. Aerogenerador síncrono multipolar de velocidad variable *y* 1.5 MW de potencia: TWT1500.*Rev. Iberoamer. Autom. Informát.* **2004**, *1*, 53–64.

7 Bianchi, F.D.; Battista, H.D.; Mantz, R.J. Wind turbine control systems. In *Principles, Modelling and Gain Scheduling Design*; Springer-Verlag: London, UK, 2007.

8 Geyler, M.; Caselitz, P. Robust multivariable pitch control design for load reduction on large wind turbines. *J. Sol. Energy Eng.* **2008**, *130*, 12.

9 Fleming, P.A.; van Wingerden, J.W.; Wright, A.D. *Comparing state-space multivariable controls to multi-siso controls for load reduction of drivetrain-coupled modes on wind turbines through field-testing*; NREL/CP-5000-53500; NREL: Colorado, CO, USA, 2011.

10 Jonkman, J.; Butterfield, S.; Musial, W.; Scott, G. *Definition of a 5 MW Reference Wind Turbine for Offshore System Development*;

NREL/TP-500-38060; Technical Report for NREL: Colorado, CO, USA; February; 3806.

11 Upwind Home page. Available online: http://www. upwind. eu (accessed on 12 December 2011).

12 Bossanyi, E.A. The design of closed loop controllers for wind turbines. *Wind Energy* **2000**, *3*, 149–163.

13 Bossanyi, E.A. Controller for 5 MW reference turbine. *European Upwind Project Report*; Garrad Hassan & Partners Ltd.: Bristol, UK, 2009. Available online: http://www.upwind.eu (accessed on 12 December 2011).

14 Wright, A.D.; Fingersh, L.J. *Advanced Control Design for Wind Turbines Part I: Control Design, Implementation, and Initial Tests*; NREL/TP-500-42437; NREL: Colorado, CO, USA, 2008.

15 Bossanyi, E.A. Wind turbine control for load reduction. *Wind Energy* **2003**, *6*, 229–244.

16 Van der Hooft, E.L.; Schaak, P.; van Engelen, T.G. *Wind Turbine Control Algorithms*; DOWEC-F1W1-EH-03094/0; Technical Report for ECN: Petten, The Netherlands, 2003.

17 Wright, A.D.; Balas, M.J. Design of controls to attenuate loads in the controls advanced research turbine. *J. Sol. Energy Eng.* **2004**, *126*, 1083.

18 Hau, M. Promising Load Estimation Methodologies for Wind Turbine Components. Technical Report for European Upwind Project. ISET: Kassel, Germany, 2009. Available online: http: //www.upwind.eu (accessed on 12 December 2012).

19 Bossanyi, E.A.; Ramtharan, G.; Savini, B. The importance of control in wind turbines design and loading. In Proceedings of the 17th Mediterranean Conference on Control & Automation, Thessaloniki, Greece, June 2009; pp. 24–26.

20 Balas, G.; Chiang, R.; Packard, A.; Safonov, M. MATLAB robust control toolbox. In *Getting Started Guide*; The MathWorks, Inc.: Natick, MA, USA, 2009.

21 Doyle, J.C.; Francis, B.A.; Tannenbaum, A.R. *Feedback Control Theory*; MacMillan: Toronto, Canada, 1992.

22 Bossanyi, E.A. *Bladed User Manual*; Garrad Hassan & Partners Ltd.: Bristol, UK, 2009

23 Frandsen, S.T. Turbulence and Turbulence Generated Structural Loading in Wind Turbine Clusters. Ph.D. Thesis, Technical University of Denmark, Roskilde, Denmark, 2007.

24 Söker, H.; Kaufeld, N. Introducing low cycle fatigue in IEC standard range pair spectra. In Proceeding of 7th German Wind Energy Conference, Wilhelmshaven, Germany, October 2004; pp. 20–21.

25 MATLAB Rainflow Counting Algorithm Toolbox. Available online: http://www.mathworks.com/MATLABcentral/fileexchange /3026 (accessed on 12 December 2011).

CHAPTER 9

$\mathcal{H}_\infty$ Control of Uncertain Fuzzy Networked Control Systems with State Quantization

Magdi S. Mahmoud

Systems Engineering Department, King Fahd University of Petroleum and Minerals, Dhahran, Saudi Arabia

ABSTRACT

The problem of robust $\mathcal{H}_\infty$ control for uncertain discrete-time Takagi and Sugeno (T-S) fuzzy networked control systems (NCSs) is investigated in this paper subject to state quantization. By taking into consideration network induced delays and packet dropouts, an improved model of network-based control is developed. A less conservative delay-dependent stability condition for the closed NCSs is derived by employing a fuzzy Lyapunov-Krasovskii functional. Robust $\mathcal{H}_\infty$ fuzzy controller is constructed that guarantee asymptotic stabilization of the NCSs and expressed in LMIbased conditions. A numerical example illustrates the effectiveness of the developed technique.

Keywords: Networked $\mathcal{H}_\infty$ Control; Fuzzy Systems; Discrete Time-Varying Delay; Linear Matrix Inequality (LMI)

1. INTRODUCTION

Fuzzy system models have been widely adopted to represent certain classes of nonlinear dynamic systems following the T-S fuzzy model [1]. Since then there have been several approaches for the study of stability analysis and robust controller synthesis using the so-called parallel distributed compensation (PDC) method for uncertain nonlinear systems [2,3]. Sufficient conditions have been derived based on the feasibility testing of a linear matrix inequality (LMI) in [4-7] and extended for classes of nonlinear discrete-time systems with time delays in [8-10] via different approaches. Recently, much attention has been paid to the stability issue of network based control systems [11]. Several results pertaining to the analysis and design of networked control systems (NCSs) enhanced their wide benefits such as reducing system wiring, ease of system diagnosis and maintenance, and increasing system agility, to name a few. However, communication network in the control loops gave rise to some new issues, especially the intermittent losses or delays of the communicated information due to use of a network, which imposes a challenge to system analysis and design. To address this challenge, many results have been developed in consideration of network-induced delay and packet dropout [12-18], with focus on stability analysis and controller design with random delays.

Further consideration of the communication of the NCSs over the channel emphasized the importance of signal quantization, which has significant impact on the performance of NCSs. In this regard, the problem of guaranteed cost control and quantized controller design were discussed in [17] by using two quantizers in the network both from sensor to controller and from controller to actuator, and the network-induced delay and data dropped were considered as well.

Recent advances converted the quantized feedback design problem into a robust control problem with sector bound uncertainties, [11] and [16-18].

Parallel investigations to the class of switched discrete-time systems with interval time-delays were developed in [19-23].

Despite the potential of these developments, the problem of how to analyze the stability of nonlinear NCSs with data drops still open. On the other hand, most industrial plants have severe nonlinearities, which lead to additional difficulties for the analysis and design of control systems. Though some issues on nonlinear NCSs have been investigated [23,24], limited work has been found on robust $\mathcal{H}_\infty$ state feedback controller design of networks for fuzzy systems with consideration of both network conditions and signal quantization.

The guaranteed cost networked control and robust $\mathcal{H}_\infty$ problem based on the T-S fuzzy model was treated in [25]. The results were derived by using a single Lyapunov function (SLF) method, which in general leas to a conservative result. Designing fuzzy controllers for a class of nonlinear networked control systems was considered in [26-28] by solving approximate uncertain linear networked Takagi-Sugeno (T-S) models with both network induced-delay and packet dropout. However, they do not quantize the signals. The foregoing facts motivate the present study.

In this research work, we address the robust $\mathcal{H}_\infty$ state feedback control problem for discrete-time networked systems with state quantization and disturbances. The T-S fuzzy systems with norm-bounded uncertainties are utilized to characterize the nonlinear NCSs. Since the computation available is often limited, the quantized feedback controller is designed under consideration of effect of network-induced delay and data dropout, the employed quantizer is time-varying. By using a new fuzzy Lyapunov-Krasovskii functional (LKF), we provide a sufficient LMI-based condition for the existence of a fuzzy controller. A numerical example shows the feasibility of the developed technique.

Notations and facts: In the sequel, the Euclidean norm is used for vectors. We use W^t and W^{-1} to denote the transpose and the inverse of any square matrix W, respectively. We use $W > 0$ $(\geq, <, \leq 0)$ to denote a symmetric positive definite (positive semi-definite, negative, negative semi-definite matrix W and I to denote the $n \times n$ identity matrix. Matrices, if their dimensions are not explicitly stated, are assumed to be compatible for algebraic operations. In symmetric block matrices or complex matrix expressions, we use the symbol ($\bullet$) to represent a term that is induced by symmetry.

Fact 1: For any real matrices Σ_1, Σ_2 and Σ_3 with appropriate dimensions and $\Sigma_3' \Sigma_3 \le I$, it follows that

$$\Sigma_1 \Sigma_3 \Sigma_2 + \Sigma_2' \Sigma_3' \Sigma_1' \le \upsilon^{-1} \Sigma_1 \Sigma_1' + \upsilon^{-1} \Sigma_2' \Sigma_2, \forall \upsilon > 0$$

Sometimes, the arguments of a function will be omitted when no confusion can arise.

2. PROBLEM DESCRIPTION

A typical networked control system typically has a clockdriven sampler and a quantizer, controller, a zero-order hold (ZOH) which is event-driven. The sampling period is assumed to be $T \ge 0$ with the sampling instants as $S_k, k = 1, \cdots, \infty$. The plant belongs to class of uncertain discrete-time systems where the parametric uncertainties are norm-bounded.

In what follows, we consider that this class is represented by Takagi-Sugeno fuzzy model composed of a set of fuzzy implications, and each implication is expressed by a linear system model. The jth rule of this TakagiSugeno model has the following form:

Rule j : If $\theta_1(k)$ is $M_{j1}, \cdots$ and $\theta_n(k)$ is M_{jn}, then

$$x(k+1) = A_{j\Delta} x(k) + B_{j\Delta} u(k) + \Gamma_{j\Delta} w(k),$$

$$y(k) = C_{j\Delta} x(k) + D_{j\Delta} u(k) + \Phi_{j\Delta} w(k),$$

$$x(k) = \psi(k), k \in [-\tau_M, 0], j = 1, 2, \cdots, r \tag{1}$$

where $\theta_1(k), \theta_2(k), \cdots, \theta_n(k)$ are the premise variables, each $M_{jm}\,(m=1,2,\cdots n)$ are the fuzzy sets, r is the number of if-then rules and $x(k)\in\Re^n$ is the state vector, $u(k)\in\Re^m$ is the control input, $y(k)\in\Re^q$ is the output, $w(k)\in\Re^p$ is the disturbance input which belongs to $\ell_2[0,\infty)$ and τ_M indicates the maximum allowable signal transmission delay. The uncertain matrices $A_{j\Delta},\cdots,\Gamma_{j\Delta}$ are represented by:

$$\begin{bmatrix} A_{j\Delta} & B_{j\Delta} & \Gamma_{j\Delta} \\ C_{j\Delta} & D_{j\Delta} & \Phi_{j\Delta} \end{bmatrix} = \begin{bmatrix} A_j & B_j & \Gamma_j \\ C_j & D_j & \Phi_j \end{bmatrix} + \begin{bmatrix} \Delta A_j & \Delta B_j & \Delta\Gamma_j \\ \Delta C_j & \Delta D_j & \Delta\Phi_j \end{bmatrix}$$

$$\begin{bmatrix} \Delta A_j & \Delta B_j & \Delta\Gamma_j \\ \Delta C_j & \Delta D_j & \Delta\Phi_j \end{bmatrix} = \begin{bmatrix} M_{1j} \\ M_{2j} \end{bmatrix} F_j(k) \begin{bmatrix} N_{1j} & N_{2j} & N_{3j} \end{bmatrix} \tag{2}$$

where the matrices $A_j, B_j, \cdots, \Phi_j$ describe the nominal dynamics and $M_{1j}, M_{2j}, N_{1j}, N_{2j}, N_{3j}$ are known constant real matrices with appropriate dimensions. The matrices $F_j(k)$ are unknown time-varying and satisfying $F_j^t(k)F_j(k), \leq I, \forall k \geq 0$.

Using a center average defuzzifier [1], product inference, and incorporating fuzzy "blending", the fuzzy system under consideration can be cast into the form

$$x(k+1) = \sum_{j=1}^{r} f_j(\theta(k))\left[A_{j\Delta}x(k) + B_{j\Delta}u(k) + \Gamma_{j\Delta}w(k)\right]$$

$$y(k) = \sum_{j=1}^{r} f_j(\theta(k))\left[C_{j\Delta}x(k) + D_{j\Delta}u(k) + \Phi_{j\Delta}w(k)\right] \tag{3}$$

Where

$$f_j(\theta(k)) = \frac{F_j(\theta(k))}{\sum_{j=1}^{r} F_j(\theta(k))},$$

$$F_j(\theta(k)) = \prod_{m=1}^{n} M_{jm}(\theta_m(k)) \tag{4}$$

where $M_{jm}(\theta_m(k))$ is the grade of membership of $\theta_m(k)$ in M_{jm}. In the sequel, we assume that

$$F_j(\theta(k)) \geq 0,\ j = 1,2,\cdots,r,\ \sum_{j=1}^{r} F_j(\theta(k)) > 0,\ \forall k \geq 0$$

and therefore

$$f_j(\theta(k)) \geq 0,\ \sum_{j=1}^{r} f_j(\theta(k)) = 1,\ \forall k \geq 0$$

Our objective in this paper is to design a fuzzy $\mathcal{H}_\infty$ state feedback controller with state quantization.

3. CONTROLLED FUZZY SYSTEM

In what follows, we proceed to consider establish the main result for the uncertain discrete-time fuzzy networked control systems described by (3) and design the quantized fuzzy $\mathcal{H}_\infty$ state feedback controller. We consider a limited capacity communication channel and for reducing the amount of data rate of transmitting in the network, which led to the increase quality of service of the network, we assume that the state vector $x(k)$ is measurable. The state signal from sensor to the controller is quantized via a quantizer, and then transmitted with a single packet. To reflect realty, network-induced time delay is modeled as an input delay and the packet dropout will be considered.

3.1. State-Feedback Control

In effect, we seek to design the state-feedback controller:

$$u(k)=g(v(k)),\quad v(k)=K\,x(d_kT) \tag{5}$$

where $g(v(k))$ is the feedback law to be defined in the sequel and $d_k,(k=1,2,3,\cdots)$ are some integers such that $\{d_1,d_2,d_3,\cdots\}\subset\{1,2,3,\cdots\}$. Introduce $\tau(k)=k-d_kT$ which contains the information of packet dropouts and improper packet sequence in the control signal. Note that $d_kT=k-(k-d_kT)=k-\tau(k)$.

It has been pointed out in [19] that when $d_{k+1}-d_k=1,$ there would be no packets dropout and the case $d_{k+1}-d_k>1$ represents continuous packets lost. In addition, when $d_{k+1}<d_k,$ the new packet reaches the destination before the old one. This case might lead to a less conservative result. In the sequel, we assume that $d_{k+1}>d_k$ and it is readily seen that

$$\tau(k)\le(d_{k+1}-d_k)T+\tau(k+1),$$
$$k\in\left[(d_kT+\tau(k)),\ (d_{k+1}T+\tau(k+1))\right]$$

It should be observed that $\tau(k)$ accounts for the time from the instant $d_k T$ when sensor nodes sample the sensor data from the plant to the instant when actuator transfer data to the plant. Extending on this, we remark that

$$\bigcup_{k=1}^{\infty}\left[\left(d_k T+\tau(k)\right),\left(d_{k+1}T+\tau(k+1)\right)\right]=\left[k_0,\infty\right),\ k_0\geq 0$$

Consequently, we define

$$\tau_m \leq \tau(k) \leq \tau_M,\ \tau_m=\eta_m T,\ \tau_M=\eta_M T$$

where $\eta_m > 0, \eta_M > 0$ are known finite integers.

3.2. Quantizer

Let the quantizer be described as

$$q(v)=\left[q_1(v_1)\, q_2(v_2)\cdots q_m(v_m)\right]',\ q_j(-v_j)=-q_j(v_j)$$

where $q(\cdot), j=1,2,\cdots,m$ is a symmetric, static and and time-invariant quantizer and the associated set of quantization levels is expressed as

$$Q=\left\{\pm\sigma_j, j=\pm 1,\pm 2,\cdots\right\}\cup\{0\} \tag{6}$$

Note that the quantization regions are quite arbitrary. In case of logarithmic quantizer, the set of quantization levels Q becomes

$$Q_{\alpha} = \{\pm\sigma_j, \sigma_j = \alpha_j \sigma_0, j = \pm 1, \pm 2, \cdots\} \cup \{\pm\sigma_0\} \cup \{0\}$$

where σ_0 is the initial state of the quantizer and $0 < \alpha_j < 1$ is a parameter associated with the quantizer $f(\cdot)$. In this regard, a particular characterization of the quantizer is given by

$$q(x) = \begin{cases} \sigma_j & \text{if } \frac{1}{1+\delta_j}\sigma_j < x \le \frac{1}{1-\delta_j}\sigma_j, x > 0 \\ 0 & \text{if } x = 0 \\ -q(-x) & \text{if } x < 0 \end{cases}$$

where $\delta_j = \frac{1-\alpha_j}{1+\alpha_j}$. It follows from [19] that, for any $q(\cdot)$, a sector bound expression can be expressed as:

$$q_j(x_j) = \left(1 + D_{q_j}(x_j)\right)x_j, \ \left|D_{q_j}(x_j)\right| \le \delta_j$$

For simplicity in exposition, we use D_q to denote $D_{q_j}(x_j)$. Thus, $q(\cdot)$ can be written as

$$q(x) = (I + D_q)x$$

We assume henceforth that the updating signal at the instant k has experienced signal transmission delay $\tau(k)$, however the delay between the sensor and quantizer is neglected. In view of the limited capacity in communication channel, the state signal from sensor to the controller is quantized via a logarithmic quantizer $q(\cdot)$ for reducing the amount of data rate of transmitting in the network. When the static and time-invariant quantizer $q(x)=x$, the state feedback controller would be in the form of $u(k)=K\,x(d_k T)$, which is the same as a traditional one.

Incorporating the notion of parallel distributed compensation, the following fuzzy state-feedback stabilizing control law is used:

Rule j: If $\theta_1(k)$ is $M_{j1},\cdots$ and $\theta_n(k)$ is M_{jn}, then

$$u(k)=K_j\left(I+D_q\right)x\left(k-\tau(k)\right) \tag{7}$$

where K_j is the control gain for rule $j,\ j=1,2,\cdots,r$. Accordingly, the overall fuzzy control law is expressed by

$$u(k)=\sum_{j=1}^{r} f_j\left(\theta\left(k-\tau(k)\right)\right) K_j\left(I+D_q\right)x\left(k-\tau(k)\right) \tag{8}$$

Applying controller (8) to system (3) with some mathematical manipulations, the resulting closed-loop system can be cast into the form:

$$x(k+1)=\sum_{j=1}^{r} f_j\left(\theta\left(k-\tau(k)\right)\right)\left[\hat{A}_{j\Delta}(k)x(k)\right.$$
$$\left.+\hat{B}_{j\Delta}(k)K_j(I+D_q)x\left(k-\tau(k)\right)+\hat{\Gamma}_{j\Delta}(k)w(k)\right],$$

$$y(k)=\sum_{j=1}^{r}f_j\big(\theta(k-\tau(k))\big)\big[\hat{C}_{j\Delta}(k)x(k)$$
$$+\hat{D}_{j\Delta}(k)K_j\big(I+D_q\big)x\big(k-\tau(k)\big)+\hat{\Phi}_{j\Delta}(k)w(k)\big] \tag{9}$$

which belongs to the class of switched time-delay system [15], where

$$\hat{A}_{j\Delta}(k)=\sum_{j=1}^{r}f_j\big(\theta(k)\big)A_{j\Delta},\hat{B}_{j\Delta}(k)=\sum_{j=1}^{r}f_j\big(\theta(k)\big)B_{j\Delta},$$

$$\hat{C}_{j\Delta}(k)=\sum_{j=1}^{r}f_j\big(\theta(k)\big)C_{j\Delta},\hat{D}_{j\Delta}(k)=\sum_{j=1}^{r}f_j\big(\theta(k)\big)D_{j\Delta}, \tag{10}$$

$$\hat{\Gamma}_{j\Delta}(k)=\sum_{j=1}^{r}f_j\big(\theta(k)\big)\Gamma_{j\Delta},\hat{\Phi}_{j\Delta}(k)=\sum_{j=1}^{r}f_j\big(\theta(k)\big)\Phi_{j\Delta}$$

4. QUANTIZED FUZZY CONTROL DESIGN

In this section, we seek to establish a sufficient condition for the solvability of the robust $\mathcal{H}_\infty$ control problem. This condition will be expressed in an LMI framework to facilitate the design of the desired fuzzy state feedback controllers. Based on the so-called parallel distributed compensation scheme, the following theorem establishes a delay-dependent stabilization condition for the closedloop fuzzy networked control system (9):

Theorem 4.1 Consider system (9). Given the bounds τ_m, τ_M and a scalar constant $\gamma > 0$, there exists a fuzzy controller in the form of (8), such that the uncertain closedloop fuzzy system (9) with an $\mathcal{H}_\infty$ disturbance attention level γ is asymptotically stable, if there exist

matrices $0<P_j, 0<Q_j, 0<Z_j, 0<S_j, 0<R_{aj}, 0<R_{cj}, K_j$,
matrices $\Psi_a, \Psi_c, \Upsilon_a, \Upsilon_c, \Theta_a, \Theta_c$ and scalars $\varepsilon_{1j}>0, \varepsilon_{2j}>0, \varepsilon_{3j}>0, \varepsilon_{4j}$, satisfying

$$\tilde{\hat{\Xi}}_j = \begin{bmatrix} \tilde{\hat{\Xi}}_{tj} & \tilde{\hat{\Xi}}_{sj} \\ \bullet & -\tilde{\hat{\Xi}}_{rj} \end{bmatrix} < 0,\ 1 \le j \le r \tag{11}$$

$$\tilde{\hat{\Xi}}_{tj} = \hat{\Xi}_{oj} + (\sigma_{1j}+\sigma_{2j}+\sigma_{3j}+\sigma_{4j})\Lambda_j \Lambda'_j$$
$$\tilde{\hat{\Xi}}_{rj} = diag\begin{bmatrix} \varepsilon_{1j}I & \varepsilon_{2j}I & \varepsilon_{3j}I & \varepsilon_{4j}I \end{bmatrix},$$

$$\tilde{\hat{\Xi}}_{sj} = \begin{bmatrix} \sigma_{1j} & \sigma_{2j} & \sigma_{3j} & \sigma_{4j} \end{bmatrix}, \tag{12}$$

$$\Lambda'_j = \begin{bmatrix} N_{1j} & N_{2j}K_j & 0 & 0 & N_{3j} & 0 & 0 & 0 & 0 & 0 \end{bmatrix}$$

$$\sigma_{1j} = \begin{bmatrix} 0 & 0 & 0 & 0 & 0 & M'_{1j} & 0 & 0 & 0 & 0 \end{bmatrix}$$

$$\sigma_{2j} = \sqrt{\tau_s}\begin{bmatrix} 0 & 0 & 0 & 0 & 0 & 0 & \sqrt{\tau_s}M'_{1j}R_{aj} & 0 & 0 & 0 \end{bmatrix}$$

$$\sigma_{3j} = \sqrt{\tau_M}\begin{bmatrix} 0 & 0 & 0 & 0 & 0 & 0 & 0 & \sqrt{\tau_M}M'_{1j}R_{cj} & 0 & 0 \end{bmatrix}$$

$$\sigma_{4j} = \sqrt{\tau_M}\begin{bmatrix} 0 & 0 & 0 & 0 & 0 & 0 & 0 & 0 & M'_{2j} & 0 \end{bmatrix} \tag{13}$$

Proof: In what follows, we adopt a parameterdependent approach [15]. Consider system (9) with $1 \le j, m \le r$ and define

$$P(k) = \sum_{j=1}^{r} f_j(\theta(k))P_j,\ Q(k) = \sum_{j=1}^{r} f_j(\theta(k))Q_j,$$

$$Z(k)=\sum_{j=1}^{r} f_j(\theta(k))Z_j,\ S(k)=\sum_{j=1}^{r} f_j(\theta(k))S_j,$$

$$R_a(k)=\sum_{j=1}^{r} f_j(\theta(k))R_{aj},\ R_c(k)=\sum_{j=1}^{r} f_j(\theta(k))R_{cj} \tag{16}$$

$$\hat{\Xi}_{aj}=\begin{bmatrix} \Xi_{ooj} & \Xi_{oaj} & \Psi_a & -\Phi_a & 0 & A_j' & \sqrt{\tau_s}(A_j-I)' & \sqrt{\tau_M}(A_j-I)' & C_j' & 0 \\ \bullet & -\Xi_{aa} & \Psi_c & -\Phi_c & 0 & K_j'B_j' & \sqrt{\tau_s}K_j'B_j' & \sqrt{\tau_M}K_j'B_j' & K_j'D_j' & 0 \\ \bullet & \bullet & -Z_j & 0 & 0 & 0 & 0 & 0 & 0 & 0 \\ \bullet & \bullet & \bullet & -S_j & 0 & 0 & 0 & 0 & 0 & 0 \\ \bullet & \bullet & \bullet & \bullet & -\gamma^2 I & \Gamma_j' & 0 & 0 & \Phi_j' & 0 \\ \bullet & \bullet & \bullet & \bullet & \bullet & -P_j+2I & 0 & 0 & 0 & B_jK_j \\ \bullet & \bullet & \bullet & \bullet & \bullet & \bullet & -R_{aj}+2I & 0 & 0 & B_jK_j \\ \bullet & \bullet & \bullet & \bullet & \bullet & \bullet & \bullet & -R_{cj}+2I & 0 & B_jK_j \\ \bullet & \bullet & \bullet & \bullet & \bullet & \bullet & \bullet & \bullet & -I & D_jK_j \\ \bullet & \bullet & \bullet & \bullet & \bullet & \bullet & \bullet & \bullet & \bullet & -I \end{bmatrix} \tag{14}$$

$$\begin{aligned} &\Xi_{ooj}=-P_j+(\tau_s+1)Q_j+Z_j+S_j+\Theta_a+\Theta_a',\ \Xi_{oaj}=-\Theta_a+\Theta_c'+\Upsilon_a-\Psi_a \\ &\Xi_{aaj}=Q_j+\Theta_c+\Theta_c'-\Upsilon_c-\Upsilon_c'+\Psi_c+\Psi_c' \end{aligned} \tag{15}$$

where $0<P_j, 0<Q_j, 0<Z_j, 0<S_j, 0<R_{aj}, 0<R_{cj}$ are matrices of appropriate dimensions and $P(k), Q(k), Z(k), S(k), R_a(k), R_c(k)$ are fuzzy weighting matrices that are directly include the membership functions instead of a single matrix, a fact that aims at relaxing the conservatism. For simplicity in notation, we let

$$A_\Delta(k)=\sum_{j=1}^{r} f_j(\theta(k-\tau(k)))\hat{A}_{j\Delta}(k),$$

$$B_\Delta(k)=\sum_{j=1}^{r} f_j(\theta(k-\tau(k)))\hat{B}_{j\Delta}(k)K_j(I+D_q),$$

$$C_\Delta(k)=\sum_{j=1}^{r} f_j(\theta(k-\tau(k)))\hat{C}_{j\Delta}(k),$$

$$D_\Delta(k)=\sum_{j=1}^{r} f_j(\theta(k-\tau(k)))\hat{D}_{j\Delta}(k)K_j(I+D_q),$$

$$\tilde{\Gamma}_{\Delta}(k)=\sum_{j=1}^{r} f_j\big(\theta(k-\tau(k))\big)\hat{\Gamma}_{j\Delta}(k),$$

$$\tilde{\Phi}_{\Delta}(k)=\sum_{j=1}^{r} f_j\big(\theta(k-\tau(k))\big)\hat{\Phi}_{j\Delta}(k) \tag{17}$$

In terms of the state increment $\delta x(k)=x(k+1)-x(k)$ and the time-span $\tau_s=\tau_M-\tau_m$ we consider the Lyapunov-Krasovskii functional (LKF):

$$V(k)=V_o(k)+V_a(k)+V_c(k)+V_m(k)+V_n(k)$$

$$V_o(k)=x'(k)P(k)x(k),$$

$$V_a(k)=\sum_{\ell=k-\tau(k)}^{k-1} x'(\ell)Q(\ell)x(\ell),$$

$$V_c(k)=\sum_{\ell=k-\tau_m}^{k-1} x'(\ell)Z\ell x(\ell)+\sum_{\ell=k-\tau_M}^{k-1} x'(\ell)S(\ell)x(\ell),$$

$$V_m(k)=\sum_{j=-\tau_M+1}^{-\tau_m}\sum_{m=k+j}^{k-1} x'(m)Q(m)x(m)$$

$$\begin{aligned}V_n(k)=&\sum_{j=-\tau_M}^{-\tau_m-1}\sum_{m=k+j}^{k-1}\delta x'(m)R_a(m)\delta x(m)\\&+\sum_{j=-\tau_M}^{-1}\sum_{m=k+j}^{k-1}\delta x'(m)R_c(m)\delta x(m)\end{aligned} \tag{18}$$

We focus initially on the case $1\le j\le r$. A straightforward computation gives the first-difference of $\Delta V(k)=V(k+1)-V(k)$ along the solutions of (17) with the help of (9) and (10) as:

$$\begin{aligned}\Delta V_o(k)&=x'(k+1)P(k+1)x(k+1)-x'(k)P(k)x(k)\\&=\big[A_\Delta(k)x(k)+B_\Delta(k)x(k-\tau(k))+\tilde{\Gamma}_\Delta(k)w(k)\big]'P(k)\big[A_\Delta(k)x(k)+B_\Delta(k)x(k-\tau(k))+\tilde{\Gamma}_\Delta(k)w(k)\big]\\&\quad-x'(k)P(k)x(k),\end{aligned}$$

$$\Delta V_a(k) \le x^t(k)Q(k)x(k) - x^t(k-\tau(k))Q(k)x(k-\tau(k)) + \sum_{j=k-\tau_M+1}^{k-\tau_m} x_j^t Q(k) x_j$$

$$\Delta V_c(k) = x^t(k)Z(k)x(k) - x^t(k-\tau_m)Z(k)x(k-\tau_m) + x^t(k)Sx(k) - x^t(k-\tau_M)S(k)x(k-d\tau_M)$$

$$\Delta V_m(k) = (\tau_M - \tau_m)x^t(k)Q(k)x(k) - \sum_{j=k-\tau_M+1}^{k-\tau_m} x^t(k)Q(k)x(k)$$

$$\Delta V_n(k) = (\tau_M - \tau_m)\delta x^t(k)R_a(k)\delta x(k) + \tau_M \delta x^t(k)R_c(k)\delta x(k) - \sum_{j=k-\tau_M}^{k-\tau_m-1} \delta x_j^t R_a(k)\delta x_j - \sum_{j=k-\tau_M}^{k-1} \delta x_j^t R_c(k)\delta x_j$$

(19)

To facilitate the delay-dependence analysis, we invoke the following identities

$$\left[2x^t(k)\Theta_a + 2x^t(k-d(k))\Theta_c\right]\left[x(k) - x(k-\tau(k)) - \sum_{j=k-\tau(k)}^{k-1} \delta x_j\right] = 0$$

$$\left[2x^t(k)\Upsilon_a + 2x^t(k-\tau(k))\Upsilon_c][x(k-\tau(k)) - x(k-\tau_M) - \sum_{j=k-\tau_M}^{k-\tau(k)-1} \delta x_j\right] = 0$$

$$\left[2x^t(k)\Psi_a + 2x^t(k-\tau(k))\Psi_c][x(k-\tau_m) - x(k-\tau(k)) - \sum_{j=k-\tau(k)}^{k-\tau_m-1} \delta x_j\right] = 0$$

(20)

for some matrices $\Theta_a, \cdots, \Psi_c$, and proceed to get

$$\begin{aligned}\Delta V(k) = {} & \Delta V_o(k) + \Delta V_a(k) + \Delta V_c(k) + \Delta V_m(k) + \Delta V_n(k) \\ & + \left[2x^t(k)\Theta_a + 2x^t(k-\tau(k))\Theta_c][x(k) - x(k-\tau(k)) - \sum_{j=k-\tau(k)}^{k-1} \delta x_j\right] \\ & + \left[2x^t(k)\Upsilon_a + 2x^t(k-\tau(k))\Upsilon_c][x(k-\tau(k)) - x(k-\tau_M) - \sum_{j=k-\tau_M}^{k-\tau(k)-1} \delta x_j\right] \\ & + \left[2x^t(k)\Psi_a + 2x^t(k-\tau(k))\Psi_c][x(k-\tau_m) - x(k-\tau(k)) - \sum_{j=k-\tau(k)}^{k-\tau_m-1} \delta x_j\right]\end{aligned}$$

(21)

In terms of

$$\xi^t(k) = \left[x^t(k)\ x^t(k-\tau(k))x^t(k-\tau_m)x^t(k-\tau_M)\right]$$

we cast (31) with $w(k)\equiv 0$ into the form:

$$\Delta V(k) \le \xi^t(k)\Pi_{\Delta j}\,\xi(k)$$

$$\Pi_{\Delta j} = \Pi_{oj} + \Xi_{1j}^t P_j \Xi_{1j} + \Xi_{2j}^t R_{aj} \Xi_{2j} + \Xi_{3j}^t R_{cj} \Xi_{3j}$$

$$\Pi_{oj} = \begin{bmatrix} \Xi_{ooj} & \Xi_{oaj} & \Psi_a & -\Upsilon_a \\ \bullet & -\Xi_{aaj} & \Psi_c & -\Upsilon_c \\ \bullet & \bullet & -Z & 0 \\ \bullet & \bullet & \bullet & -S \end{bmatrix},$$

$$\Xi_{1j}^t = \begin{bmatrix} A_\Delta^t \\ B_\Delta^t \\ 0 \\ 0 \end{bmatrix}, \Xi_{3j}^t = \begin{bmatrix} \sqrt{\tau_M}\,(A_\Delta - I)^t \\ \sqrt{\tau_M}\,B_\Delta \\ 0 \\ 0 \end{bmatrix},$$

$$\Xi_{2j}^t = \begin{bmatrix} \sqrt{\tau_s}\,(A_\Delta - I)^t \\ \sqrt{\tau_s}\,B_\Delta \\ 0 \\ 0 \end{bmatrix} \tag{22}$$

where $\Xi_{oo}, \Xi_{aa}, \Xi_{oa}$ are given by (15). If $\Pi_{\Delta j} < 0$ for all admissible uncertainties satisfying (2), then by Schur complements it follows from (32) that $\Delta V(k) < 0,$ for any $\xi(k) \neq 0$ guaranteeing the internal stability.

Proceeding further and to assure the closed-loop stability with γ-disturbance attenuation, we follow [15] to get:

$$\begin{aligned}\Delta V(k)+y'(k)y(k)-\gamma^2 w'(k)w(k) &= \Delta V(k)-\gamma^2 w'(k)w(k)\\ &+\left[C_\Delta(k)x(k)+D_\Delta(k)x(k-\tau(k))+\tilde{\Phi}_\Delta(k)w(k)\right]'\\ &\left[C_\Delta(k)x(k)+D_\Delta(k)x(k-\tau(k))+\tilde{\Phi}_\Delta(k)w(k)\right]\Delta V(k)\\ &\leq\left[\xi(k)\,w'(k)\right]\Sigma_{\Delta j}(k)\left[\xi'(k)\,w'(k)\right]'<0\end{aligned} \tag{23}$$

when $\Sigma(k)<0$ where

$$\begin{aligned}&\Sigma_{\Delta j}(k)=\Xi_{oj}+\Xi'_{aj}P_j\Xi_{aj}+\Xi'_{cj}R_{aj}\Xi_{cj}+\Xi'_{ej}R_{cj}\Xi_{ej}+\Xi'_{mj}\Xi_{mj}\\ &\Xi_{oj}=\begin{bmatrix}\Pi_{oj} & 0\\ \bullet & -\gamma^2 I\end{bmatrix},\ \Xi'_{aj}=\begin{bmatrix}\Xi'_{1j}\\ \tilde{\Gamma}'_\Delta(k)\end{bmatrix},\ \Xi'_{cj}=\begin{bmatrix}\Xi'_{2j}\\ 0\end{bmatrix},\\ &\Xi_{mj}=\left[C'_\Delta(k)\ \ D'_\Delta(k)\ \ 0\ \ 0\ \ \tilde{\Phi}'_\Delta(k)\right]',\\ &\Xi'_{ej}=\begin{bmatrix}\Xi'_{3j}\\ 0\end{bmatrix}\end{aligned} \tag{24}$$

Next, by applying Fact 1, we obtain

$$\begin{aligned}\Xi_{\Delta j}\leq\;&\breve{\Xi}_{oj}+\varepsilon_{1j}^{-1}\hat{\sigma}_{1j}\hat{\sigma}'_{1j}+\varepsilon_{2j}^{-1}\hat{\sigma}_{2j}\hat{\sigma}'_{2j}\\ &+\varepsilon_{3j}^{-1}\hat{\sigma}_{3j}\hat{\sigma}'_{3j}+\varepsilon_{4j}^{-1}\hat{\sigma}_{4j}\hat{\sigma}'_{4j}\\ &<0\end{aligned} \tag{25}$$

for some scalars $\varepsilon_{1j} > 0, \varepsilon_{2j} > 0, \varepsilon_{3j} > 0, \varepsilon_{4j} > 0$. Note that

$f_j(\theta(k)) \geq 0, f_j(\theta(k-\tau(k))) \geq 0, 1 \leq j \leq r.$

The quantities $\hat{\sigma}_{1j}, \hat{\sigma}_{2j}, \hat{\sigma}_{3j}, \hat{\sigma}_{4j}$ correspond to $\sigma_{1j}, \sigma_{2j}, \sigma_{3j}, \sigma_{4j}$ given by (13) after deleting the last element, and

$$\breve{\Xi}_{oj} = \begin{bmatrix} \Xi_{ooj} & \Xi_{oaj} & \Psi_a & -\Phi_a & 0 & A_j' & \sqrt{\tau_s}\left(A_j - I\right)' & \sqrt{\tau_M}\left(A_j - I\right)' & C_j' \\ \bullet & -\Xi_{aaj} & \Psi_c & -\Phi_c & 0 & \Omega_{aj}' & \sqrt{\tau_s}\Omega_{aj}' & \sqrt{\tau_M}\Omega_{aj}' & \Omega_{cj}' \\ \bullet & \bullet & -Z & 0 & 0 & 0 & 0 & 0 & 0 \\ \bullet & \bullet & \bullet & -S & 0 & 0 & 0 & 0 & 0 \\ \bullet & \bullet & \bullet & \bullet & -\gamma^2 I & \Gamma_j' & 0 & 0 & \Phi_j' \\ \bullet & \bullet & \bullet & \bullet & \bullet & -P_j^{-1} & 0 & 0 & 0 \\ \bullet & \bullet & \bullet & \bullet & \bullet & \bullet & -R_{aj}^{-1} & 0 & 0 \\ \bullet & \bullet & \bullet & \bullet & \bullet & \bullet & \bullet & -R_{cj}^{-1} & 0 \\ \bullet & \bullet & \bullet & \bullet & \bullet & \bullet & \bullet & \bullet & -I \end{bmatrix},$$

$$\Omega_{aj} = \left(B_j K_j \left(I + D_q\right)\right), \ \Omega_{cj} = \left(D_j K_j \left(I + D_q\right)\right) \tag{26}$$

where $\Xi_{ooj}, \Xi_{aaj}, \Xi_{oaj}$ are given by (15). Further convexification of $\Xi_{\Delta j}$ in (35) yields

$$\Xi_{\Delta j} \leq \hat{\Xi}_{oj} + \varepsilon_{1j}^{-1}\sigma_{1j}\sigma_{1j}' + \varepsilon_{2j}^{-1}\sigma_{2j}\sigma_{2j}' + \varepsilon_{3j}^{-1}\sigma_{3j}\sigma_{3j}' + \varepsilon_{4j}^{-1}\sigma_{4j}\sigma_{4j}' < 0 \tag{27}$$

By Schur complements using the algebraic inequality $(X - I)X^{-1}(X - I) \geq 0$ for any matrix $X > 0$, the desired stability condition can then be cast into the LMI (11), which concludes the proof.

Remark 4.1 It is significant to observe that Theorem 4.1 provides a delay-dependent condition for the design of robust $\mathcal{H}_\infty$ for fuzzy NCS in terms of feasibility testing of a family of strict LMIs with a total number of LMIvariables as $6(r+1)+r$. The key feature is that the matrix gain K_j is treated as a direct LMI variable. This will eventually lessen the conservatism in robust fuzzy control design.

Remark 4.2 It is worthy to note that the number of LMIs increases linearly with the number of rules r which limits the applicability of the method for very large values of r. Had we used

$$P(k)=\sum_{j=1}^{r} f_j(\theta(k))P_j := P,\ Q(k)=\sum_{j=1}^{r} f_j(\theta(k))P_j := Q,$$

$$Z(k)=\sum_{j=1}^{r} f_j(\theta(k))P_j := Z, S(k)=\sum_{j=1}^{r} f_j(\theta(k))P_j := S,$$

$$R_a(k)=\sum_{j=1}^{r} f_j(\theta(k))R_{aj} := R_a, R_c(k)=\sum_{j=1}^{r} f_j(\theta(k))R_{cj} := R_a$$

then Theorem 4.1 reduces to the following corollary:

Corollary 4.1 Given the bounds τ_m, τ_M and a scalar constants $\gamma > 0$, there exists a fuzzy controller in the form of (8), such that the uncertain closed-loop fuzzy system (9) with an $\mathcal{H}_\infty$ disturbance attention level γ is asymptotically stable, if there exist matrices $0<P,$

$0<Q, 0<Z, 0<S, 0<R_a, 0<R_c$, matrices $K_j, \Psi_a, \Psi_c, \Upsilon_a, \Upsilon_c, \Theta_a, \Theta_c$ and scalars $\varepsilon_{1j}>0, \varepsilon_{2j}>0, \varepsilon_{3j}>0, \varepsilon_{4j}$, satisfying

$$\tilde{\Xi} = \begin{bmatrix} \tilde{\Xi}_t & \tilde{\Xi}_s \\ \bullet & -\tilde{\Xi}_r \end{bmatrix} < 0, \tag{28}$$

$$\tilde{\Xi}_t = \tilde{\Xi}_o + (\sigma_{1j} + \sigma_{2j} + \sigma_{3j} + \sigma_{4j})\Lambda\Lambda'$$

$$\tilde{\Xi}_r = diag\begin{bmatrix} \varepsilon_{1j}I & \varepsilon_{2j}I & \varepsilon_{3j}I & \varepsilon_{4j}I \end{bmatrix},$$

$$\tilde{\Pi}_s = \begin{bmatrix} \sigma_1 & \sigma_2 & \sigma_3 & \sigma_4 \end{bmatrix},$$

$$\Lambda'_j = \begin{bmatrix} N_{1j} & N_{2j}K_j & 0 & 0 & N_{3j} & 0 & 0 & 0 & 0 & 0 \end{bmatrix}$$

$$\sigma_{1j} = \begin{bmatrix} 0 & 0 & 0 & 0 & 0 & M'_{1j} & 0 & 0 & 0 & 0 \end{bmatrix}$$

$$\sigma_{2j} = \sqrt{\tau_s}\begin{bmatrix} 0 & 0 & 0 & 0 & 0 & 0 & \sqrt{\tau_s}M'_{1j}R_{aj} & 0 & 0 & 0 \end{bmatrix}$$

$$\sigma_{3j} = \sqrt{\tau_M}\begin{bmatrix} 0 & 0 & 0 & 0 & 0 & 0 & 0 & \sqrt{\tau_M}M'_{1j}R_{cj} & 0 & 0 \end{bmatrix},$$

$$\sigma_{4j} = \sqrt{\tau_M}\begin{bmatrix} 0 & 0 & 0 & 0 & 0 & 0 & 0 & 0 & M'_{2j} & 0 \end{bmatrix} \tag{29}$$

$$\hat{\Xi}_o = \begin{bmatrix} \Xi_{oo} & \Xi_{oa} & \Psi_a & -\Phi_a & 0 & A'_j & \sqrt{\tau_s}(A_j - I)' & \sqrt{\tau_M}(A_j - I)' & C'_j & 0 \\ \bullet & -\Xi_{aa} & \Psi_c & -\Phi_c & 0 & K'_jB'_j & \sqrt{\tau_s}K'_jB'_j & \sqrt{\tau_M}K'_jB'_j & K'_jD'_j & 0 \\ \bullet & \bullet & -Z & 0 & 0 & 0 & 0 & 0 & 0 & 0 \\ \bullet & \bullet & \bullet & -S & 0 & 0 & 0 & 0 & 0 & 0 \\ \bullet & \bullet & \bullet & \bullet & -\gamma^2 I & \Gamma'_j & 0 & 0 & \Phi'_j & 0 \\ \bullet & \bullet & \bullet & \bullet & \bullet & -P+2I & 0 & 0 & 0 & B_jK_j \\ \bullet & \bullet & \bullet & \bullet & \bullet & \bullet & -R_a+2I & 0 & 0 & B_jK_j \\ \bullet & \bullet & \bullet & \bullet & \bullet & \bullet & \bullet & -R_c+2I & 0 & B_jK_j \\ \bullet & \bullet & \bullet & \bullet & \bullet & \bullet & \bullet & \bullet & -I & D_jK_j \\ \bullet & \bullet & \bullet & \bullet & \bullet & \bullet & \bullet & \bullet & \bullet & -I \end{bmatrix} \tag{30}$$

$$\Xi_{ooj} = -P + (\tau_s + 1)Q + Z + S + \Theta_a + \Theta'_a$$

$$\Xi_{aaj} = Q + \Theta_c + \Theta'_c - \Upsilon_c - \Upsilon'_c + \Psi_c + \Psi'_c,$$

$$\Xi_{oaj} = -\Theta_a + \Theta'_c + \Upsilon_a - \Psi_a \tag{31}$$

and the number of LMI variables would be $12 + r$. The price paid is that the LKF becomes non-fuzzy.

5. SPECIAL CASES

In this section, we seek to derive a sufficient condition for the solvability of the robust $\mathcal{H}_\infty$ control problem for NCS without quantizer.

NCS without Quantizer

In this case, the resulting closed-loop fuzzy system can be expressed as:

$$x(k+1)=\sum_{j=1}^{r} f_j(\theta(k))f_j(\theta(k))\big[A_{j\Delta}(k)x(k) +B_{j\Delta}(k)K_j(I+D_q)x(k-\tau(k))+\Gamma_{j\Delta}(k)w(k)\big],$$

$$\begin{aligned}y(k)=\sum_{j=1}^{r} f_j(\theta(k))f_j(\theta(k))\big[&C_{j\Delta}(k)x(k)\\ &+D_{j\Delta}(k)K_j(I+D_q)x(k-\tau(k))\\ &+\Phi_{j\Delta}(k)w(k)\big]\end{aligned} \quad (32)$$

The corresponding control design is given by the following corollary:

Corollary 5.1 Given the bounds τ_m, τ_M and a scalar constants $\gamma > 0$, there exists a fuzzy controller in the form of (8), such that the uncertain closed-loop fuzzy system (23)

with an $\mathcal{H}_\infty$ disturbance attention level γ is asymptotically stable, if there exist matrices $0<P_j, 0<Q_j, 0<Z_j$, $0<S_j, 0<R_{aj}, 0<R_{cj}$, matrices $K_j, \Psi_a, \Psi_c, \Upsilon_a, \Upsilon_c, \Theta_a$, Θ_c and scalars $\varepsilon_{1j}>0, \varepsilon_{2j}>0, \varepsilon_{3j}>0, \varepsilon_{4j}$: satisfying

$$\tilde{\Xi}_j=\begin{bmatrix}\overline{\Xi}_{tj} & \tilde{\Xi}_{sj}\\ \bullet & -\tilde{\Xi}_{rj}\end{bmatrix}<0,\ 1\le j\le r \quad (33)$$

$$\bar{\Xi}_{ij} = \bar{\Xi}_{oj} + \left(\sigma_{1j} + \sigma_{2j} + \sigma_{3j} + \sigma_{4j}\right)\Lambda_j \Lambda_j'$$

$$\bar{\Xi}_{oj} = \begin{bmatrix} \Xi_{ooj} & \Xi_{oaj} & \Psi_a & -\Phi_a & 0 & A_j' & \sqrt{\tau_s}\left(A_j - I\right)' & \sqrt{\tau_M}\left(A_j - I\right)' & C_j' \\ \bullet & -\Xi_{aa} & \Psi_c & -\Phi_c & 0 & K_j' B_j' & \sqrt{\tau_s} K_j' B_j' & \sqrt{\tau_M} K_j' B_j' & K_j' D_j' \\ \bullet & \bullet & -Z_j & 0 & 0 & 0 & 0 & 0 & 0 \\ \bullet & \bullet & \bullet & -S_j & 0 & 0 & 0 & 0 & 0 \\ \bullet & \bullet & \bullet & \bullet & -\gamma^2 I & \Gamma_j' & 0 & 0 & \Phi_j' \\ \bullet & \bullet & \bullet & \bullet & \bullet & -P_j + 2I & -R_{aj} + 2I & 0 & 0 \\ \bullet & \bullet & \bullet & \bullet & \bullet & \bullet & -R_{aj} + 2I & 0 & 0 \\ \bullet & \bullet & \bullet & \bullet & \bullet & \bullet & \bullet & -R_{cj} + 2I & 0 \\ \bullet & \bullet & \bullet & \bullet & \bullet & \bullet & \bullet & \bullet & -I \end{bmatrix} \quad (34)$$

where the various terms are as in (13)-(15).

6. EXAMPLE

In what follows, a typical simulation example is considered to illustrate the fuzzy controller design procedure developed in Theorem 4.1. A class of discrete-time fuzzy networked control systems model with state quantization is described by:

Rule 1: If $x_1(k)$ is M_1, then

$$x(k+1) = A_{1\Delta} x(k) + B_{1\Delta} u(k) + \Gamma_{1\Delta} w(k),$$

$$y(k) = C_{1\Delta} x(k) + D_{1\Delta} u(k) + \Phi_{1\Delta} w(k),$$

$$A_1 = \begin{bmatrix} -0.05 & -0.8 \\ 0.7 & -0.04 \end{bmatrix}, B_1 = \begin{bmatrix} 0.1 \\ 0.2 \end{bmatrix}, D_1 = 0.2$$

$$\Gamma_1 = \begin{bmatrix} -0.1 & -0.3 \\ 0.2 & 0.5 \end{bmatrix}, C_1' = \begin{bmatrix} 0.2 \\ 0.5 \end{bmatrix}, \Phi_1' = \begin{bmatrix} 0.1 \\ 0.4 \end{bmatrix},$$

$$M_{11}=\begin{bmatrix}0.2\\0.2\end{bmatrix}, N_{11}^{\prime}=\begin{bmatrix}0.1\\0.2\end{bmatrix}, N_{31}^{\prime}=\begin{bmatrix}0.1\\0.2\end{bmatrix}$$

$$M_{21}=0.2, N_{21}=0.05$$

Rule 2: If $x_2(k)$ is M_2, then

$$x(k+1)=A_{2\Delta}x(k)+B_{2\Delta}u(k)+\Gamma_{2\Delta}w(k),$$

$$y(k)=C_{2\Delta}x(k)+D_{2\Delta}u(k)+\Phi_{2\Delta}w(k),$$

$$A_2=\begin{bmatrix}-0.03 & 0.5\\-1.2 & -0.04\end{bmatrix}, B_2=\begin{bmatrix}0.09\\0.06\end{bmatrix}, D_2=0.3$$

$$\Gamma_2=\begin{bmatrix}0.2 & -0.6\\0.1 & -0.4\end{bmatrix}, C_2^{\prime}=\begin{bmatrix}0.3\\0.05\end{bmatrix}, \Phi_2^{\prime}=\begin{bmatrix}0.1\\0.4\end{bmatrix},$$

$$M_{12}=\begin{bmatrix}0.1\\0.1\end{bmatrix}, N_{12}^{\prime}=\begin{bmatrix}0.05\\0.08\end{bmatrix}, N_{32}^{\prime}=\begin{bmatrix}0.03\\0.06\end{bmatrix},$$

$$M_{22}=0.3, N_{22}=0.04$$

Rule 3: If $x_3(k)$ is M_3, then

$$x(k+1)=A_{3\Delta}x(k)+B_{3\Delta}u(k)+\Gamma_{3\Delta}w(k),$$

$$y(k)=C_{3\Delta}x(k)+D_{3\Delta}u(k)+\Phi_{3\Delta}w(k),$$

$$A_3=\begin{bmatrix}-0.05 & -0.8\\0.7 & -0.04\end{bmatrix}, B_3=\begin{bmatrix}0.07\\0.05\end{bmatrix}, D_3=0.4$$

$$\Gamma_3=\begin{bmatrix}-0.1 & -0.3\\0.2 & 0.5\end{bmatrix}, C_3^{\prime}=\begin{bmatrix}0.5\\0.04\end{bmatrix}, \Phi_3^{\prime}=\begin{bmatrix}0.08\\0.06\end{bmatrix},$$

$$M_{13}=\begin{bmatrix}0.2\\0.1\end{bmatrix}, N_{13}^{\prime}=\begin{bmatrix}0.06\\0.04\end{bmatrix}, N_{33}^{\prime}=\begin{bmatrix}0.04\\0.04\end{bmatrix},$$

$M_{23} = 0.4, N_{23} = 0.05$

The membership functions for the rules 1, 2, 3 are

$$M_1(x_1(k)) = \frac{1}{1 + exp(-3x_1(k))},$$

$$M_2(x_2(k)) = \frac{1}{1 + exp(-2x_2(k))},$$

$$M_3(x_3(k)) = 1 - M_1(x_1(k)) - M_2(x_2(k))$$

For the purpose of implementation, we consider the fuzzy system to be controlled through a network. A quantizer $q(\cdot)$ is selected to be of of logarithmic type with $\alpha_1 = 0.8, \alpha_2 = 0.7, \alpha_3 = 0.9,$ leading to $\delta_1 = 0.1111,$ $\delta_2 = 0.1765, \delta_3 = 0.0526$. The bounds on data packet dropout are selected as $\tau_m = 0.6, \tau_M = 3.2,$ respectively. Using the Matlab LMI solver, the feasible solution of Theorem 4.1 yields the fuzzy $\mathcal{H}_\infty$ state-feedback controller gains of the form:

$$\gamma = 0.25,$$
$$K_1 = [18.5403 \quad 9.3657],$$
$$K_2 = [17.7234 \quad 8.8945]$$
$$K_3 = [15.7715 \quad 8.2863]$$

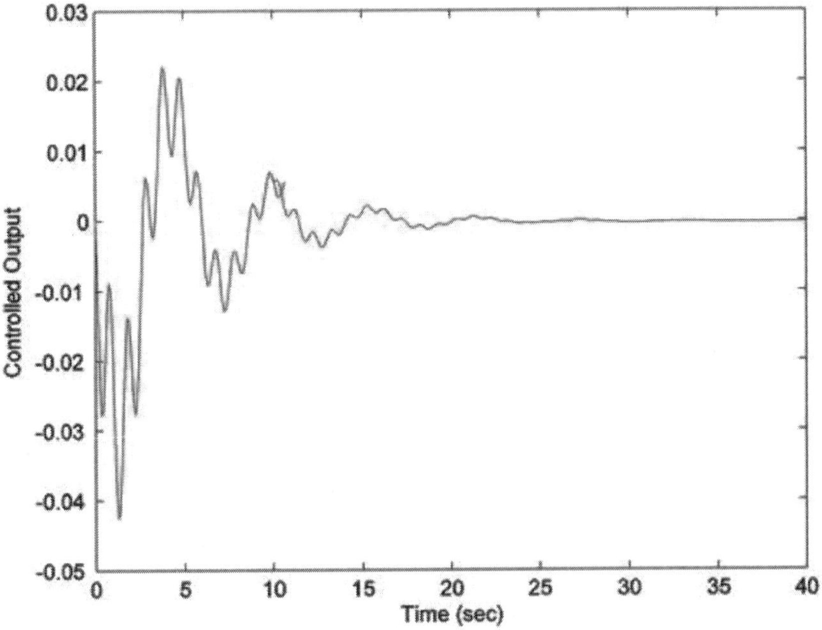

Figure 1. Controlled-output trajectory.

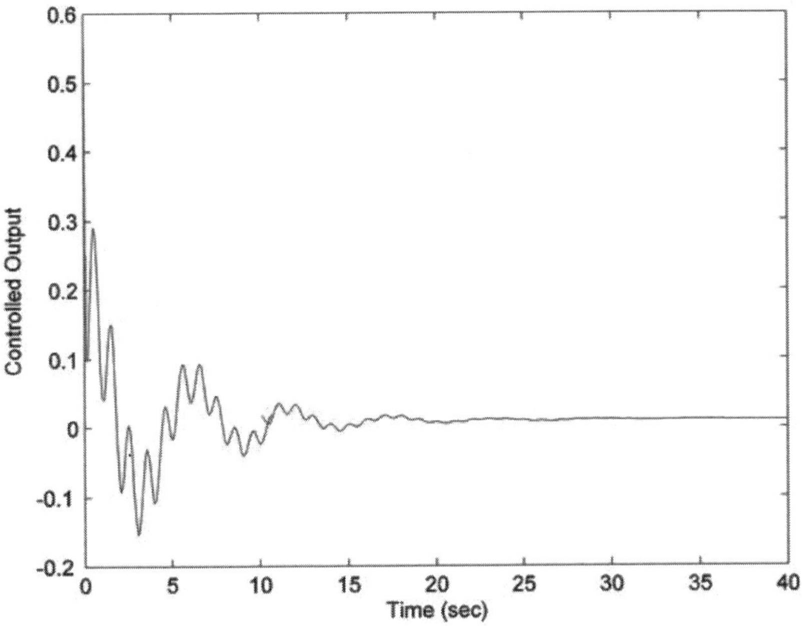

Figure 2. First state trajectory.

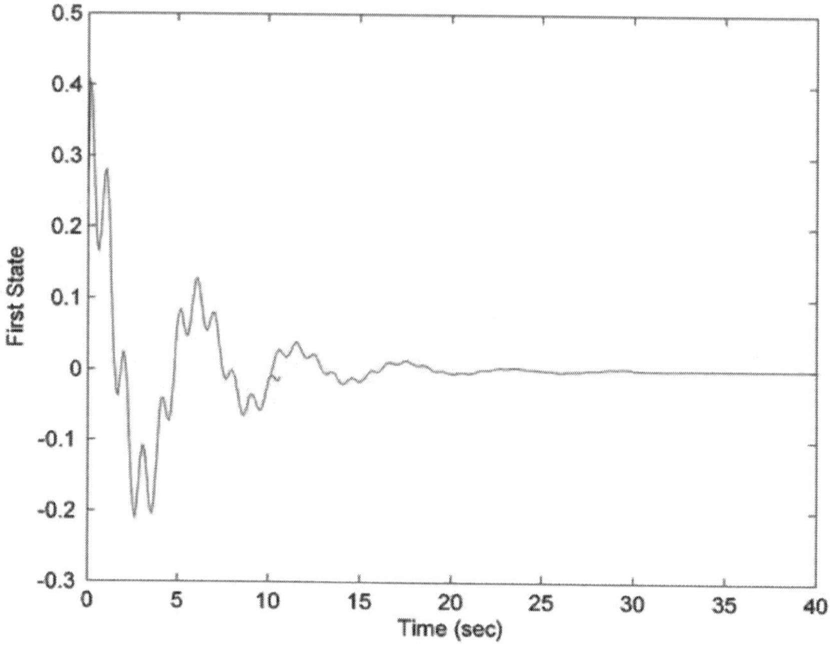

Figure 3. Second state trajectory.

The simulation results of the state and controlledoutput trajectories are plotted in Figures 1-3. It is quite evident that all the state and output variables of the fuzzy system settle at the equilibrium level within 20 sec.

7. CONCLUSION

We have addressed the problem of robust $\mathcal{H}_\infty$ statefeedback controller design for discrete-time TakagiSugeno (T-S) fuzzy networked control systems including state quantization. A quantized feedback fuzzy controller has been designed under consideration of effect of network-induced delay and data dropout, and the timevarying quantizer has been selected to be logarithmic. By employing a fuzzy Lyapunov-Krasovskii functional, we have derived some LMI-based sufficient conditions for the existence of fuzzy controller. A numerical example has been given to illustrate the efficiency of the theoretic results.

ACKNOWLEDGEMENTS

The author would like to thank the deanship of scientific research (DSR) at KFUPM for financial support through research group project RG1105-1.

REFERENCES

1 T. Takagi and M. Sugeno, "Fuzzy Identification of Systems and Its Applications to Modeling and Control," IEEE Transactions on Systems, Man, and Cybernetics, Vol. 15, No. 1, 1985, pp. 116-132.

2 K. Tanaka, T. Ikeda and H. Wang, "Robust Stabilization of a Class of Uncertain Nonlinear Systems via Fuzzy Control: Quadratic Stabilizability, $\mathcal{H}_\infty$ Control Theory, and Linear Matrix Inequalities," IEEE Transactions on Fuzzy Systems, Vol. 4, No. 1, 1996, pp. 1-13.

3 X. Guan and C. Chen, "Delay-Dependent Guaranteed Cost Control for T-S Fuzzy Systems with Time Delays," IEEE Transactions on Fuzzy Systems, Vol. 12, No. 2, 2004, pp. 236-249.

4 Y. Cao and P. Frank, "Analysis and Synthesis of Nonlinear Time-Delay systems via Fuzzy Control Approach," IEEE Transactions on Fuzzy Systems, Vol. 8, No. 2, 2000, pp. 200- 211.

5 S. Zhou and T. Li, "Robust Stabilization for Delayed Discrete-Time Fuzzy Systems via Basis-Dependent Lyapunov-Krasovskii Function," Fuzzy Sets and Systems, Vol. 151, No. 1, 2005, pp. 139-153.

6 S. Xu and J. Lam, "Robust $\mathcal{H}_\infty$ Control for Uncertain Discrete-Time-Delay Fuzzy Systems via Output Feedback Controllers," IEEE Transactions on Fuzzy Systems, Vol. 13, No. 1, 2005, pp. 82-93.

7 H. J. Lee, J. B. Park and G. Chen, "Robust Fuzzy Control of Nonlinear Systems with Parametric Uncertainties," IEEE Transactions on Fuzzy Systems, Vol. 9, No. 2, 2001, pp. 369-379.

8 S. Chen, W. Chang and S. Su, "Robust Static OutputFeedback Stabilization for Nonlinear Discrete-Time Systems with Time Delay via Fuzzy Control Approach," Fuzzy Sets and Systems, Vol. 13, No. 2, 2005, pp. 263- 272.

9 Y. Cao and P. Frank, "Robust $\mathcal{H}_\infty$ Disturbance Attenuation for a Class of Uncertain Discrete-Time Fuzzy Systems," IEEE Transactions on Fuzzy Systems, Vol. 8, No. 4, 2000, pp. 406-415.

10 H. Wu, "Delay-dependent Stability Analysis and Stabilization for Discrete-Time Fuzzy Systems with State Delay: A Fuzzy Lyapunov-Krasovskii Functional Approach," IEEE Transactions on Fuzzy Systems, Vol. 36, No. 4, 2006, pp. 954-962.

11 H. Gao, T. Chen and J. Lam, "A New Delay System Approach to Network-Based Control," Automatica, Vol. 44, No. 1, 2008, pp. 39-52.

12 J. Wu and T. Chen, "Design of Networked Control Systems with Packet Dropouts," IEEE Transactions on Automatic Control, Vol. 52, No. 7, 2007, pp. 1314-1319.

13 A. Zhang and L. Yu, "Output Feedback Stabilization of Networked Control Systems with Packet Dropouts," IEEE Transactions on Automatic Control, Vol. 52, No. 9, 2007, pp. 1705-1710.

14 L. Zhang, Y. Shi, T. Chen and B. Huang, "A New Method for Stabilization of Networked Control Systems with Random Delays," IEEE Transactions on Automatic Control, Vol. 50, No. 8, 2005, pp. 1177-1181.

15 M. S. Mahmoud, "Switched Time-Delay Systems," Springer-Verlag, New York, 2010.

16 E. Tian, D. Yue and C. Peng, "Quantized Output Feedback Control for Networked Control Systems," Information Sciences, Vol. 178, 2008, pp. 2734-2749.

17 D. Yue, C. Peng and G. Tang, "Guaranteed Cost Control of Linear Systems over Networks with State and Input Quantizations," IET

Control Theory and Application, Vol. 153, No. 6, 2006, pp. 658-664.

18 P. Chen and C. Yu, "Networked $\mathcal{H}_\infty$ Control of Linear Systems with State Quantization," Information Sciences, Vol. 177, 2007, pp. 5763-5774.

19 M. S. Mahmoud, "Delay-Dependent $\mathcal{H}_\infty$ Filtering of a Class of Switched Discrete-Time State-Delay Systems," Signal Processing, Vol. 88, No. 11, 2008, pp. 2709-2719.

20 M. S. Mahmoud, A. W. Saif and P. Shi, "Stabilization of Linear Switched Delay Systems: $\mathcal{H}_2$ and $\mathcal{H}_2$ Methods," Journal of Optimization Theory and Applications, Vol. 142, No. 3, 2009, pp. 583-607.

21 J. Zhang, Y. Xia and M. S. Mahmoud, "Robust Generalized H_2 and $\mathcal{H}_\infty$ Static Output Feedback Control for Uncertain Discrete-Time Fuzzy Systems," IET Control Theory and Applications, Vol. 3, No. 7, 2009, pp. 865-876.

22 H. Gao, P. Shi and J. Wang, "Parameter-Dependent Robust Stability of Uncertain Time-Delay Systems," Computational & Applied Mathematics, Vol. 206, No. 1, 2007, pp. 366-373.

23 H. Gao, J. Lam, C. Wang and Y. Wang, "Delay-Dependent Output-Feedback Stabilization of Discrete-Time Systems with Time-Varying State Delay," IEE Proceedings of Control Theory and Applications, Vol. 151, No. 6, 2004, pp. 691-698.

24 H. Gao and T. Chen, "A New Approach to Quantized Feedback Control Systems," Automatica, Vol. 44, No. 2, 2008, pp. 534-542.

25 G. Walsh and H. Ye, "Scheduling of Networked Control Systems," IEEE Control Systems Magazine, Vol. 21, No. 1, 2001, pp. 57-65.

26 H. Zhang, D. Yang and T. Chai, "Guaranteed Cost Networked Control for T-S Fuzzy Systems with Time Delays," Applications and Reviews, Vol. 37, No. 2, 2007, pp. 160-172.

27 L. Zhang and H. Fang, "Fuzzy Controller Design for Networked Control System with Time-Variant Delays," Journal of Systems Engineering and Electronics, Vol. 17, No. 1, 2006, pp. 172-176

28 X. Jiang and Q. Han, "On Designing Fuzzy Controllers for a Class of Nonlinear Networked Control Systems," IEEE Transactions on Fuzzy Systems, Vol. 16, No. 4, 2008, pp. 1050-1060.

Index